Analysis and Control of Age-Dependent Population Dynamics

MATHEMATICAL MODELLING:
Theory and Applications

VOLUME 11

This series is aimed at publishing work dealing with the definition, development and application of fundamental theory and methodology, computational and algorithmic implementations and comprehensive empirical studies in mathematical modelling. Work on new mathematics inspired by the construction of mathematical models, combining theory and experiment and furthering the understanding of the systems being modelled are particularly welcomed.

Manuscripts to be considered for publication lie within the following, non-exhaustive list of areas: mathematical modelling in engineering, industrial mathematics, control theory, operations research, decision theory, economic modelling, mathematical programmering, mathematical system theory, geophysical sciences, climate modelling, environmental processes, mathematical modelling in psychology, political science, sociology and behavioural sciences, mathematical biology, mathematical ecology, image processing, computer vision, artificial intelligence, fuzzy systems, and approximate reasoning, genetic algorithms, neural networks, expert systems, pattern recognition, clustering, chaos and fractals.

Original monographs, comprehensive surveys as well as edited collections will be considered for publication.

The titles published in this series are listed at the end of this volume.

Analysis and Control of Age-Dependent Population Dynamics

by

Sebastian Aniţa
Faculty of Mathematics,
University 'Al. I. Cuza' Iaşi

and

Institute of Mathematics,
Romanian Academy,
Iaşi, Romania

KLUWER ACADEMIC PUBLISHERS
DORDRECHT / BOSTON / LONDON

Library of Congress Cataloging-in-Publication Data

Anita, Sebastian.
Analysis and control of age-dependent population dynamics / by Sebastian Anita.
p. cm. -- (Mathematical modelling--theory and applications ; v. 11)
Includes bibliographical references (p.).

1. Age-structured populations--Mathematical models. 2. Population biology--Mathematical models. I. Title. II. Series.

QH352 .A54 2000
577.8'8'015118--dc21

00-064699

ISBN 978-90-481-5590-3

Published by Kluwer Academic Publishers,
P.O. Box 17, 3300 AA Dordrecht, The Netherlands.

Sold and distributed in North, Central and South America
by Kluwer Academic Publishers,
101 Philip Drive, Norwell, MA 02061, U.S.A.

In all other countries, sold and distributed
by Kluwer Academic Publishers,
P.O. Box 322, 3300 AH Dordrecht, The Netherlands.

Printed on acid-free paper

CONTENTS

PREFACE

The material of the present book is an extension of a graduate course given by the author at the University "Al.I. Cuza" Iaşi and is intended for students and researchers interested in the applications of optimal control and in mathematical biology.

Age is one of the most important parameters in the evolution of a biological population. Even if for a very long period age structure has been considered only in demography, nowadays it is fundamental in epidemiology and ecology too.

This is the first book devoted to the control of continuous age structured population dynamics. It focuses on the basic properties of the solutions and on the control of age structured population dynamics with or without diffusion.

The main goal of this work is to familiarize the reader with the most important problems, approaches and results in the mathematical theory of age-dependent models. Special attention is given to optimal harvesting and to exact controllability problems, which are very important from the economical or ecological points of view. We use some new concepts and techniques in modern control theory such as Clarke's generalized gradient, Ekeland's variational principle, and Carleman estimates. The methods and techniques we use can be applied to other control problems.

The book is intended to be accesible to graduate students whose mathematical background includes basic courses in functional analysis, ordinary and partial differential equations. Some technical details of the proofs are omitted but the reader can always fill the gaps. Several exercises are proposed at the end of the chapters. For most of them we have introduced substantial hints.

The references are not complete and refer only to the works closely related or used in this book.

I would like to express my gratitude to Professor Viorel Barbu for his support and interest in the development of the present work.

I enjoyed the fruitful cooperation and discussions with Professor Mimmo Iannelli. I extend to him my warmest thanks.

Thanks are also due to Professor Michel Langlais for stimulating discussions which helped me to improve this material.

NOTATIONS

$\mathbf{R} = (-\infty, +\infty)$ – the set of real numbers
$\mathbf{R}^+ = [0, +\infty)$
$r^+ = \max\{r, 0\}, r^- = -\min\{r, 0\} = r^+ - r$
$\mathbf{R}^n$ – the n–dimensional Euclidean space
$\mathbf{C}$ – the set of complex numbers
$\Omega \subset \mathbf{R}^n$ – an open subset of $\mathbf{R}^n$; $\partial\Omega$ is the boundary of Ω
$\overline{\Omega}$ – the closure of the set Ω
$\omega \subset\subset \Omega$ means that $\overline{\omega}$ is a compact subset of Ω

$\nabla u = \left(\frac{\partial u}{\partial x_1}, \frac{\partial u}{\partial x_2}, ..., \frac{\partial u}{\partial x_n}\right)$ – the gradient of u

$x \cdot y = \sum_{i=1}^n x_i y_i$ – the scalar product of $x = (x_1, ..., x_n)$, $y = (y_1, ..., y_n)$

$\Delta = \sum_{i=1}^{n} \frac{\partial^2}{\partial x_i^2}$ – the Laplace operator

$\| \cdot \|_X$ – the norm of space X

$Du(a, t) = \lim_{\varepsilon \to 0} \frac{u(a + \varepsilon, t + \varepsilon) - u(a, t)}{\varepsilon}$

$Du(a, t, x) = \lim_{\varepsilon \to 0} \frac{u(a + \varepsilon, t + \varepsilon, x) - u(a, t, x)}{\varepsilon}$

$\frac{\partial u}{\partial \nu}$ – the outward normal derivative

$L^p(\Omega) = \{u$ measurable on Ω; $\int_\Omega |u|^p dx < +\infty\}$, $1 \leq p < +\infty$

$H^m(\Omega)$ – the Sobolev space $\{u \in L^2(\Omega);\ D^\alpha u \in L^2(\Omega),\ |\alpha| \leq m\}$
$C^k([0, T]; X)$ – the space of continuously differentiable functions $u : [0, T] \to X$, of order up to and including k

$L^p(0, T; X) = \{u$ measurable from $(0, T)$ to X; $\int_0^T \|u(t)\|_X^p dt < +\infty\}$

CHAPTER 1

INTRODUCTION

One of the most challenging problems in science is to model biological phenomena. The great number of parameters involved in the dynamics of a biological population makes deduction of a general model quite difficult. An early concern in this matter was to find such a general model, but the relevance of some parameters appeared only over time and allowed improvement of the models. The purpose of this chapter is to describe some of the most important continuous models and basic aspects of population dynamics. We will deal with the case of a single population living in a habitat, all of its individuals of a certain age being perfectly equal (in particular it is assumed that there are no sex differences).

1.1. Overview of the problems

FIRST MODELS

The simplest population dynamics model was introduced by T.R. Malthus in 1798 in a famous treatise [68]. He supposed that the rate of population growth is proportional to the size of the population. More precisely, if we denote by $P(t)$ the total number of individuals of a population at time t, then the following differential equation is satisfied,

$$P'(t) = \alpha P(t), \quad t > 0, \tag{1.1}$$

where $\alpha \in \mathbf{R}$ is the intrinsic growth constant (so-called malthusian parameter). Usually α is taken as the difference between the fertility and mortality rates.

The solution of (1.1) is

$$P(t) = P(0)e^{\alpha t}, \quad t \geq 0$$

and led to the conclusion of exponential growth (for $\alpha > 0$), exponential decay (for $\alpha < 0$) or stationary state (for $\alpha = 0$) of the population.

Malthus considered that the fertility rate is greater than the mortality rate ($\alpha > 0$) and so he concluded the "famous" exponential population growth. In reality no biological population can grow beyond some intrinsically determined level.

The malthusian model describes a single homogeneous population with all individuals of the population identical. It also assumes that the population lives isolated in an invariant habitat with no limit to its resources. This model does not take into account the effects of crowding and of limitations on resources.

An improved model, which seems to be more realistic, was proposed by P.F. Verhulst in 1838 (see [83]). Here the total population P satisfies

$$P'(t) = \alpha P(t) - \gamma P^2(t), \quad t > 0, \tag{1.2}$$

where $\alpha \in \mathbf{R}$ is the intrinsic growth constant, while α/γ is called the environmental carrying capacity ($\gamma > 0$). Overcrowding and the limitation of resources lead to the appearence of an additional mortality rate, which is proportional to the total population ($\gamma P(t)$). Any solution of (1.2) tends to a constant as $t \to +\infty$.

The models of Malthus and Verhulst are examples of continuous population models. One of the deficiencies of the above ordinary differential equation models is that they do not take into account an age structure which, in many situations can influence population size and behaviour.

AGE-STRUCTURED MODELS WITHOUT DIFFUSION

The theory of continuous population dynamics has been developed by many mathematicians. One of the most important improvements was the consideration of age distribution within population.

In 1911, F.R. Sharpe and A. Lotka [80], and in 1926 A.G. McKendrick [70] proposed two of the first age-structured continuous models.

Consider a single population species and denote by $p(a,t)$ the density of individuals of age a at the moment t.

In the above mentioned models (of Sharpe-Lotka-McKendrick) the population dynamics is described by the equation

$$Dp(a,t) + \mu(a)p(a,t) = 0, \quad a \in (0, a_\dagger),\ t > 0, \tag{1.3}$$

where $a_\dagger$ is the maximal age for the population species and $\mu(a)$ is the mortality rate and depends only on age a. By Dp is denoted the directional derivative of p with respect to the direction $(1,1)$, i.e.,

$$Dp(a,t) = \lim_{\varepsilon \to 0} \frac{p(a+\varepsilon, t+\varepsilon) - p(a,t)}{\varepsilon}.$$

For p smooth enough this derivative is equal to $\dfrac{\partial p}{\partial a}(a,t) + \dfrac{\partial p}{\partial t}(a,t)$.

The birth process is described by the so-called birth law

$$p(0,t) = \int_0^{a_\dagger} \beta(a)p(a,t)da, \quad t > 0, \tag{1.4}$$

where $p(0,t)$ is the number of newborn population at the moment t and $\beta(a)$ is the fertility rate and depends only on age a. Here $\beta(a)p(a,t)$ gives the density of newborns at the moment t, with parents of age a.

The initial age distribution of the population is

$$p(a,0) = p_0(a), \quad a \in (0, a_\dagger), \tag{1.5}$$

where p_0 is a known function.

Equations (1.3) – (1.5) constitute the classical linear model of age dependent population dynamics.

The analysis of (1.3) – (1.5) led to a similar conclusion as for the malthusian model, as regards the large time behaviour of the solution. It has been proved that only in rare cases the solution tends to a nontrivial equilibrium state for $t \to +\infty$. In fact, in real situations this often happens.

It has been observed that in real situations the fertility and mortality rates also depend on the total population $P(t)$,

$$P(t) = \int_0^{a_\dagger} p(a,t)da, \quad t \geq 0,$$

as in the Verhulst model (see [51]). Due to the overcrowding effect and to limitation of resources, the mortality μ is an increasing function of P and the fertility is a decreasing function of P.

These considerations lead to non-linear continuous age-dependent models, which are more realistic. Such models were proposed by M.E. Gurtin and R.C. MacCamy [41] in 1974 and by F. Hoppensteadt [50] in 1975. The equation proposed for the population dynamics is

$$Dp(a,t) + \mu(a,t,P(t))p(a,t) = 0, \quad a \in (0,a_\dagger),\ t > 0, \tag{1.6}$$

and the birth process is described by the nonlocal boundary condition

$$p(0,t) = \int_0^{a_\dagger} \beta(a,t,P(t))p(a,t)da, \quad t > 0\ . \tag{1.7}$$

The significance of p, β, μ, $a_\dagger$, Dp (in (1.6) and (1.7)) is the same as before. The fertility and mortality rates are now functions of three independent variables, which is why these nonlinear age-dependent models of population dynamics provide a biologically more realistic description of the behaviour of the population.

The books of F. Hoppensteadt [50], B. Charlesworth [30], J. Metz and O. Diekmann [72] and J. Murray [76] provide a survey of the applicability of age-structured models. The linear model of the growth of an age-structured population is a basic mathematical tool in demography. The monographs by N. Keyfitz [54], [55] contain the fundamentals of the theory and some applications to demographic data.

AGE-STRUCTURED MODELS WITH DIFFUSION

Another phenomenon which is important in population dynamics is spatial diffusion. M.E. Gurtin [40] (in 1973) took into consideration diffusion for age-structured populations. This model was extended by M.E. Gurtin and R.C. MacCamy [42] (in 1977), [44] (in 1981). The Gurtin-MacCamy model describes a biological population which is free to move in a habitat $\Omega \subset \mathbf{R}^n$, where $n \in \{1,\ 2,\ 3\}$. If we denote by $p(a,t,x)$ the density of individuals of age $a \in [0, a_\dagger]$, at the moment $t \geq 0$ and in the location $x \in \overline{\Omega}$, then the following equations are satisfied:

$$Dp(a,t,x) + \mu(a, P(t,x))p(a,t,x) - k\Delta p(a,t,x) = 0, \tag{1.8}$$

$(a,t,x) \in (0,a_\dagger) \times (0,+\infty) \times \Omega$ (where Δ denotes the Laplace operator with respect to x),

$$p(0,t,x) = \int_0^{a_\dagger} \beta(a, P(t,x))p(a,t,x)da, \quad (t,x) \in (0,+\infty) \times \Omega, \tag{1.9}$$

where $a_\dagger$, β and μ have the same significance as in the above discussed models, $k > 0$ is the diffusion constant,

$$Dp(a,t,x) = \lim_{\varepsilon \to 0} \frac{p(a+\varepsilon, t+\varepsilon, x) - p(a,t,x)}{\varepsilon},$$

and

$$P(t,x) = \int_0^{a_\dagger} p(a,t,x)da .$$

Problem (1.8) – (1.9) has been investigated under appropriate boundary conditions.

The importance of diffusion (including the case of nonlinear diffusion) in population dynamics has been emphasized by D.G. Aronson [13].

The basic properties of the solutions of age-dependent population dynamics (e.g., existence, uniqueness, positivity and large time behaviour) have been intensively studied by several authors (see the bibliographical comments in the next chapters).

CONTROL OF AGE-STRUCTURED POPULATION DYNAMICS

Population control is the process of forcing a population in order to obtain a certain behaviour of it. The intervention strategy constitutes the control.

Some of the most important control problems are the optimal harvesting of natural or farmed populations (such as plants and fish) and the control of pests and parasites.

The main attention has been given to the "optimal harvesting" problem which has been intensively investigated in the last decades. Early work on the topic considered the problem of deterministic harvesting, ignoring age structure. The first study of the problem of harvesting a discrete age-structured population is attributed to L. Lefkovitch [63], [64]. The optimal harvesting strategy for a continuous age-structured model was discussed by C. Rores and W. Fair [78] in 1980. They assume that the control strategy (harvesting effort) is time-independent. Optimal control of population dynamics is discussed in [81] - Chapter 8.

M. Gurtin and L. Murphy [46], [47] investigate optimal harvesting for the nonlinear model introduced by M. Gurtin and R.C. MacCamy in [41]. They make appropriate assumptions which allow the system of partial differential equations to be reduced to a system of ordinary differential equations. The harvesting effort is assumed to be age-independent.

This problem can be formulated as follows:

$$\textbf{Maximize} \int_0^T \int_0^\infty u(t)p^u(a,t)da\ dt,$$

subject to $u \in \mathcal{U} = \{v \in L^\infty(0,T);\ 0 \le v(t) \le L \text{ a.e. in } (0,T)\}$, where p^u is the solution of

$$\begin{cases} Dp(a,t) + \mu(a,P(t))p(a,t) = -u(t)p(a,t), & a \in (0,+\infty),\ t \in (0,T), \\ p(0,t) = \displaystyle\int_0^\infty \beta(a,P(t))p(a,t)da, & t \in (0,T), \\ p(a,0) = p_0(a), & a \in (0,+\infty), \\ P(t) = \displaystyle\int_0^\infty p(a,t)da, & t \in (0,T), \end{cases}$$

($T,\ L \in (0,+\infty)$). Here, the maximal age is $+\infty$ and the fertility and the mortality rates are independent of time t.

Function $u \in \mathcal{U}$ is the control (or "harvesting effort") and

$$\int_0^T \int_0^\infty u(t)p^u(a,t)da\ dt$$

represents the total harvest in the time interval $[0,T]$. Remark that the control u acts as a supplementary mortality rate.

The aim of the study of this problem is to find the control $u^* \in \mathcal{U}$ which realizes the maximum of the function

$$u \longmapsto \int_0^T \int_0^\infty u(t)p^u(a,t)da\ dt$$

(cost function). This control u^*, is the optimal control or the optimal harvesting effort and

$$\int_0^T \int_0^\infty u^*(t)p^{u^*}(a,t)da\ dt$$

is the optimal harvest.

A similar problem is considered by L.F. Murphy and S. Smith [75].

It is more realistic to take the harvesting effort as a function of age a and time t. M. Brokate in [24], [25] was the first to study this case (even for a more general cost function).

An optimal harvesting problem governed by the Gurtin-MacCamy model with separable mortality rate is considered by S. Aniţa [10].

The linear model with periodic vital rates and periodic inflow is treated in [11]. The optimal control is shown to be of bang-bang type.

N. Medhin [71] considers both finite and infinite horizon optimal harvesting in age-structured populations.

In the last few years, the control for age-dependent population dynamics has been intensively developed by mathematical researchers. We mention some of the main directions of this development, which will be studied in the next chapters:

- finding maximum principles and using them in order to approximate the optimal control (e.g., Brokate [24], [25]);
- approximating the optimal control via fractional step schemes (e.g., Aniţa [10]);
- the study of time-periodic harvesting problems (see Aniţa, Iannelli, Kim and Park [11]);
- the controllability problem (see Ainseba and Aniţa [3]).

In the chapters which follow it will be our objective to develop the analysis of some of the most important models and to study the most significant control problems related to age-dependent population dynamics.

1.2. General models of population dynamics with diffusion

In this section we will derive some general continuous models with diffusion for age-structured populations.

Suppose that a given single population species is free to move in an open and bounded habitat $\Omega \subset \mathbf{R}^N$ ($N \in \{1,2,3\}$) with (smooth enough) boundary $\partial\Omega$. Denote by $p(a,t,x)$ the density of individuals of age $a \in [0,a_\dagger]$

($a_\dagger \in (0,+\infty)$ is the maximal age of the population species), at time $t \geq 0$ and location $x \in \overline{\Omega}$. The integral

$$\int_{\alpha_1}^{\alpha_2} p(a,t,x)da, \quad t \geq 0, \; x \in \overline{\Omega},$$

represents the population density of age $a \in [\alpha_1, \alpha_2]$ at time t and location x, and

$$P(t,x) = \int_0^{a_\dagger} p(a,t,x)da$$

gives the total population density at time t and location x.

Assume that the diffusion of the population is governed by the Nernst law:

"The flux of population through a surface S (which is the boundary of an arbitrary open subset $V \subset\subset \Omega$) in the time interval $[t, t+dt]$ is given by

$$-k \cdot dt \int_S \frac{\partial p}{\partial \nu}(a,t,x)d\sigma,$$

where $k > 0$ is the diffusion coefficient and $\dfrac{\partial p}{\partial \nu}$ is the (outward) normal derivative to S."

The population density of age a, located in V at the moment t is

$$\int_V p(a,t,x)dx,$$

and the population density of age $a + dt$, located in V at the moment $t + dt$, is

$$\int_V p(a+dt, t+dt, x)dx \; .$$

Consider μ the mortality rate. This gives the proportion of dying individuals and depends on age a, time t, location x and $P(t,x)$, the total population density at time t and location x. The balance law shows that the number of individuals of age a, at the moment t, located in V, that die in the time interval $[t, t+dt]$ is

$$dt \int_V \mu(a,t,x,P(t,x))p(a,t,x)dx = k \cdot dt \int_S \frac{\partial p}{\partial \nu}(a,t,x)d\sigma$$
$$+ \int_V p(a,t,x)dx - \int_V p(a+dt,t+dt,x)dx \; .$$

Dividing now by dt we obtain

$$\int_V \mu(a,t,x,P(t,x))p(a,t,x)dx = -\int_V Dp(a,t,x)dx + k \int_S \frac{\partial p}{\partial \nu}(a,t,x)d\sigma \; ,$$

where

$$Dp(a,t,x) = \lim_{\varepsilon \to 0} \frac{p(a+\varepsilon, t+\varepsilon, x) - p(a,t,x)}{\varepsilon} .$$

$(= (\frac{\partial}{\partial a} + \frac{\partial}{\partial t})p(a,t,x)$ if p is smooth enough). Dp is the "growth rate".

Using Ostrogradski's formula we get

$$\int_V [Dp(a,t,x) + \mu(a,t,x,P(t,x))p(a,t,x)]\, dx = k \int_V \Delta p(a,t,x) dx .$$

Since V is an arbitrary open subset $V \subset\subset \Omega$, we may conclude that the population dynamics is described by

$$Dp(a,t,x) + \mu(a,t,x,P(t,x))p(a,t,x) - k\Delta p(a,t,x) = 0 ,$$

for $a \in (0, a_\dagger)$, $t > 0$, $x \in \Omega$.

If a certain infusion of population in the domain Ω appears (of growth rate $f(a,t,x)$), then the evolution of p is governed by

$$Dp(a,t,x) + \mu(a,t,x,P(t,x))p(a,t,x) - k\Delta p(a,t,x) = f(a,t,x) , \tag{2.1}$$

for $a \in (0, a_\dagger)$, $t > 0$, $x \in \Omega$.

The most appropriate boundary conditions for (2.1) are

$$p(a,t,x) = 0, \quad a \in (0, a_\dagger),\ t > 0,\ x \in \partial\Omega, \tag{2.2}$$

(the homogeneous Dirichlet condition) which describes the case of a *completely inhospitable* boundary;

$$\frac{\partial p}{\partial \nu}(a,t,x) = 0, \quad a \in (0, a_\dagger),\ t > 0,\ x \in \partial\Omega, \tag{2.3}$$

(the homogeneous Neumann condition) when there is no exchange of population (migration) across $\partial\Omega$;

$$\frac{\partial p}{\partial \nu}(a,t,x) + \alpha p(a,t,x) = 0, \quad a \in (0, a_\dagger),\ t > 0,\ x \in \partial\Omega \tag{2.4}$$

($\alpha > 0$) when there is a migration of population across the boundary, which is proportional to the density p on the boundary of Ω. Sometimes it is appropriate to consider a nonhomogeneous condition corresponding to (2.2), (2.3) or (2.4).

The birth process is described by the *"renewal law"*

$$p(0,t,x) = \int_0^{a_\dagger} \beta(a,t,x,P(t,x))p(a,t,x)da, \quad t > 0,\ x \in \Omega . \tag{2.5}$$

Here β is the fertility rate and gives the proportion of newborn population at moment t and location x, with parents of age a. This rate depends on age a, time t, location x and $P(t,x)$, the total population at time t and location x.

The initial density of individuals is given by

$$p(a,0,x) = p_0(a,x), \quad a \in (0,a_\dagger),\ x \in \Omega\ . \tag{2.6}$$

The usual hypotheses on β, μ, p_0 and f will be made precise in the next chapters. We wish only to emphasize that all these hypotheses will be based on observations of biological populations.

If the fertility and the mortality rates depend only on age, their graphs are as in Figure 1.

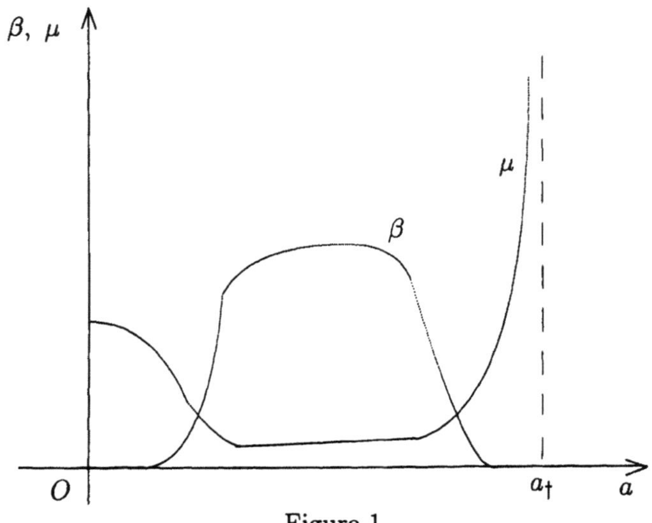

Figure 1

In general, the birth rate β is a decreasing function with respect to the fourth variable, while the mortality rate μ is increasing with respect to the fourth variable. Note that the fertility rate β is zero in a neighbourhood of $a := 0$ and in a neighbourhood of $a := a_\dagger$. On the other hand for x fixed, the following condition on μ holds,

$$\int_0^{a_\dagger} \mu(a_\dagger - s, t - s, x, 0)da = +\infty \quad \text{for all } (t,x) \in [0,+\infty) \times \overline{\Omega},$$

where $\mu(\cdot,\cdot,\cdot,0)$ is extended by 0 outside $[0,a_\dagger]\times[0,+\infty)\times\overline{\Omega}$. Under certain assumptions on μ, this condition is necessary and sufficient in order to have $a_\dagger$ as the maximal age of the population species.

Another model for population dynamics is the following:

$$Dp(a,t,x)+\mu(a,t,x,P(t,x))p(a,t,x)-k\Delta p(a,t,x)$$

$$+\eta(p(a,t,x)-\psi(a,x))\ni f(a,t,x)\ , \tag{2.7}$$

for $a\in(0,a_\dagger)$, $t>0$, $x\in\Omega$, where

$$\eta(r)=\begin{cases}0, & \text{if } r<0\\ \mathbf{R}^+, & \text{if } r=0\ .\end{cases}$$

(see Garroni and Langlais [39]). Here $\psi(a,x)\geq 0$ is a critical density. We remark that in this case $p(a,t,x)\leq\psi(a,x)$ for $a\in(0,a_\dagger)$, $t>0$, $x\in\Omega$. For $p(a,t,x)<\psi(a,x)$, the dynamics is described by an equation (because $\eta(p(a,t,x)-\psi(a,x))=0$) and for $p(a,t,x)=\psi(a,x)$, a certain additional mortality rate (due to crowding) appears and makes the population density $p(a,t,x)$ remain below the level $\psi(a,x)$.

There are also some biological populations described by models with nonlinear diffusion. This happens when the diffusion coefficient depends on the population density. The population dynamics is described in this case by

$$Dp(a,t,x)+\mu(a,t,x)p(a,t,x)-\Delta\gamma(p(a,t,x))=f(a,t,x), \tag{2.8}$$

for $a\in(0,a_\dagger)$, $t>0$, $x\in\Omega$, where $\gamma:\mathbf{R}\to\mathbf{R}$ is a function with certain properties (see Aniţa [7]).

Of course, equations (2.7) and (2.8) should be studied together with an appropriate boundary condition (e.g., (2.2), (2.3) or (2.4)), with the nonlocal boundary condition (2.5) and the initial condition (2.6).

The models introduced in this section seem to be the most realistic continuous ones for age-dependent population dynamics with diffusion.

We have to mention that each mathematical model has its applicability area and also its limitations.

1.3. Age-dependent epidemic models

It is well known that in many diseases the chronological age of the individual is an important factor of vulnerability and infectiousness. The attention paid to age-dependent epidemic models is in connection with the modeling of childhood diseases. A large class of age-dependent models is discussed by F. Hoppensteadt [50].

We shall describe an age-dependent model for epidemics, due to R. Anderson and R. May [4], K. Dietz [33] and D. Schenzle [33], [79].

Consider a biological population which is described by the model (1.3)-(1.5) in Section 1.2. Because of epidemics, the population is partitioned into three classes: *susceptibles, infectives* and *removed* (the individuals who have been infective and are now immune, dead or isolated). Denote by $s(a,t)$, $i(a,t)$, $r(a,t)$ the density of susceptibles, infectives, respectively removed of age a, at the moment t. It is obvious that

$$p(a,t) = s(a,t) + i(a,t) + r(a,t),$$

where $p(a,t)$ is the population density, corresponding to age a and moment t.

The transmission dynamics of the disease is described by the following system:

$$\begin{cases} Ds(a,t) + \mu(a)s(a,t) = -\lambda(a,t)s(a,t) + d_2(a)i(a,t), \\ Di(a,t) + \mu(a)i(a,t) = \lambda(a,t)s(a,t) - (d_1(a) + d_2(a))\,i(a,t), \\ Dr(a,t) + \mu(a)r(a,t) = d_1(a)i(a,t), \end{cases} \tag{3.1}$$

$(a,t) \in (0, a_\dagger) \times (0, +\infty)$, where

$\lambda(a,t)$	is the infection rate (the rate at which susceptibles get the disease and become infectives);
$d_1(a)$	is the removal rate (the rate at which infectives become removed);
$d_2(a)$	is the recovery rate (the rate at which infectives become susceptibles).

Together with (3.1) we must consider the conditions

$$\begin{cases} s(0,t) = \displaystyle\int_0^{a_\dagger} \beta(a)\,(s(a,t) + (1-k_1)i(a,t) + (1-k_2)r(a,t))\,da, \\ i(0,t) = k_1 \displaystyle\int_0^{a_\dagger} \beta(a)i(a,t)da, \\ r(0,t) = k_2 \displaystyle\int_0^{a_\dagger} \beta(a)r(a,t)da, \end{cases} \tag{3.2}$$

$t > 0$, where $k_1,\ k_2 \in [0,1]$ represent the vertical transmission parameters of infectiveness and immunity, respectively (and give the proportion of newborns that remain in the class of their parents).

The initial conditions are

$$\begin{cases} s(a,0) = s_0(a), \\ i(a,0) = i_0(a), \\ r(a,0) = r_0(a), \end{cases} \tag{3.3}$$

$a \in (0, a_\dagger)$. In the model (3.1) – (3.3) the vital rates β and μ are supposed to be unaffected by the disease. This model constitutes the so-called *S-I-R model.*

It is obvious that if we add the equations in (3.1), (3.2) and (3.3) we obtain (1.3), (1.4) and (1.5), respectively (with $p_0(a) = s_0(a) + i_0(a) + r_0(a)$). In the special case of a disease that does not import immunity ($d_1 \equiv 0$ and $r_0 \equiv 0$) we get the *S-I-S model.*

For a mathematical study of these two models we refer to the monograph by M. Iannelli [52] and the references therein.

We note that there are several optimal control problems related to the age-dependent epidemic models. One of the most important is the problem of "optimal vaccination".

Suppose that the effect of a certain vaccination lasts for a certain period (does not give definitive immunity) and has to be refreshed periodically. So, the population is structured by age a, number of vaccinations j and the time since the last vaccination v.

Assume that susceptibles and recovered can be infected, but the recovered have a lower susceptibility. The infected recover and begin to belong to the removed class with lifetime immunity. Denote by

$s_j(a,t,v)$	the density of susceptibles, j-times vaccinated;
$w_j(a,t,v)$	the density of partially protected, j-times vaccinated;
$r(a,t)$	the density of removed;
$u_j(a,v)$	the vaccination strategy (control) corresponding to the j destination class;
$u_0(a)$	the control for the nonvaccinated individuals;
$k(a_1,a_2)$	the contact distribution of an infected individual at age a_1 and an uninfected individual of age a_2;
$\eta(a,v)$	the rate of loss of vaccine immunity;
$d_1(a)$	the removal rate;
k_0	the probability of vertical transmission.

The dynamics is described by the following equations

$$Ds_0(a,t) = -(\mu(a)+u_0(a))\,s_0(a,t) - \beta(a)\frac{s_0(a,t)}{P(t)}\int_0^{a_\dagger} k(\alpha,a)i(\alpha,t)d\alpha,$$

$$\left(D+\frac{\partial}{\partial v}\right)s_j(a,t,v) = -(\mu(a)+u_j(a,v))\,s_j(a,t,v)$$
$$-\beta(a)\frac{s_j(a,t,v)}{P(t)}\int_0^{a_\dagger} k(\alpha,a)i(\alpha,t)d\alpha + \eta(a,v)w_j(a,t,v),$$

$$\left(D+\frac{\partial}{\partial v}\right)w_j(a,t,v) = -(\mu(a)+u_j(a,v))\,w_j(a,t,v)$$
$$-\beta(a)\frac{w_j(a,t,v)}{P(t)}\int_0^{a_\dagger} k(\alpha,a)i(\alpha,t)d\alpha - \eta(a,v)w_j(a,t,v),$$

$$Di(a,t) = -(\mu(a)+d_1(a))\,i(a,t)$$
$$+\frac{\sum_j\int_0^{a_\dagger}\beta(a)\,(s_j(a,t,v)+w_j(a,t,v))\,dv}{P(t)}\int_0^{a_\dagger} k(\alpha,a)i(\alpha,t)d\alpha,$$
$$Dr(a,t) = -\mu(a)r(a,t)+d_1(a)i(a,t).$$

Here $P(t)$ is the total population at the moment t.

The following conditions are also satisfied

$$s_0(0,t) = \int_0^{a_\dagger}[\beta(a)(r(a,t)+s_0(a,t)$$
$$+\sum_{j\geq 1}\int_0^{a_\dagger}(s_j(a,t,v)+w_j(a,t,v))dv)+k_0\beta(a)i(a,t)]da,$$

$$s_j(0,t,v)=0 \quad , \quad s_j(a,t,0)=0,$$

$$w_1(0,t,v)=0 \quad , \quad w_1(a,t,0)=u_0(a)s_0(a,t),$$

$$w_j(0,t,v) = 0, \qquad j \geq 1,$$

$$w_j(a,t,0) = \int_0^{a_\dagger} u_{j-1}(a,v)\,(s_{j-1}(a,t,v) + w_{j-1}(a,t,v))\,dv,$$

$$i(0,t) = (1-k_0)\int_0^{a_\dagger} \beta(a)i(a,t)da,$$

$$r(0,t) = 0.$$

In the case of vaccination the goal is to find a strategy (control) which minimizes a certain cost functional $\mathcal{L}(u_0, u_1, ..., s_0, s_1, ..., w_0, w_1, ..., i, r)$. So, it is very important to define an appropriate cost functional.

For a discussion of the possible cost functionals and a detailed treatment of this vaccination model we refer to the work of K.P. Hadeler [48], [49] and J. Mueller [48], [49], [74].

CHAPTER 2

ANALYSIS OF AGE-DEPENDENT POPULATION DYNAMICS

This chapter is devoted to the basic properties of the models of age-dependent population dynamics without diffusion. The main focus is on the existence, uniqueness and positivity of solutions of the linear model and of the nonlinear model. Some comparison results, which will be used later in the study of the optimal control of population dynamics, are stated. The asymptotic behaviour of the solutions is also investigated. Finally we establish some basic properties of the solution of linear periodic age-dependent population dynamics.

Throughout this chapter $a_\dagger$, $T \in (0, +\infty)$, $Q_T = (0, a_\dagger) \times (0, T)$ and Dp is given by

$$Dp(a,t) = \lim_{\varepsilon \to 0} \frac{p(a+\varepsilon, t+\varepsilon) - p(a,t)}{\varepsilon}$$

(Dp is a directional derivative).

2.1. The linear age-dependent model

The linear model which we consider here, corresponding to an idealized situation, provides a good insight into age-structured phenomena.

Consider the following linear model describing the evolution of an age-structured population:

$$\begin{cases} Dp(a,t) + \mu(a,t)p(a,t) = f(a,t), & (a,t) \in Q_T, \\ p(0,t) = \displaystyle\int_0^{a_\dagger} \beta(a,t)p(a,t)da, & t \in (0,T), \\ p(a,0) = p_0(a), & a \in (0, a_\dagger). \end{cases} \tag{1.1}$$

We choose the Banach space $L^1(0, a_\dagger)$ as the mathematical setting of our problem, because it is the most natural choice for many population problems.

Throughout this section we work under the following assumptions, excepting the case when we will explicitly mention other hypotheses:

(A1) $\beta \in L^\infty(Q_T),\ \beta(a,t) \geq 0$ a.e. $(a,t) \in Q_T$,

(A2) $\mu \in L^1_{loc}([0,a_\dagger) \times [0,T]),\ \mu(a,t) \geq 0$ a.e. $(a,t) \in Q_T$,

(A3) $p_0 \in L^1(0,a_\dagger),\ p_0(a) \geq 0$ a.e. $a \in (0,a_\dagger)$,
$f \in L^1(Q_T),\ f(a,t) \geq 0$ a.e. $(a,t) \in Q_T$.

We mention that assumptions **(A1)-(A3)** are satisfied by a large class of biological populations.

We shall first study the existence of a solution to (1.1).

By *a solution* to (1.1) we mean a function $p \in L^\infty(0,T;L^1(0,a_\dagger))$, absolutely continuous along almost every characteristic line (of equation $a - t = k$, $(a,t) \in \overline{Q}_T$, $k \in \mathbf{R}$), such that

$$\begin{cases} Dp(a,t) + \mu(a,t)p(a,t) = f(a,t) & \text{a.e. in } Q_T, \\ \lim\limits_{\varepsilon \to 0^+} p(\varepsilon, t+\varepsilon) = \int_0^{a_\dagger} \beta(a,t)p(a,t)da & \text{a.e. } t \in (0,T), \\ \lim\limits_{\varepsilon \to 0^+} p(a+\varepsilon,\varepsilon) = p_0(a) & \text{a.e. } a \in (0,a_\dagger)\ . \end{cases} \tag{1.2}$$

Since p is absolutely continuous along almost every characteristic line, relations $(1.2)_{2,3}$ are meaningful.

It is now obvious that the meaning of $p(0,t)$ and $p(a,0)$ are

$$p(0,t) = \lim_{\varepsilon \to 0^+} p(\varepsilon, t+\varepsilon) \quad \text{a.e. } t \in (0,T)$$

and

$$p(a,0) = \lim_{\varepsilon \to 0^+} p(a+\varepsilon,\varepsilon) \quad \text{a.e. } a \in (0,a_\dagger)\ .$$

Now, we can prove the existence and uniqueness of a solution for problem (1.1). Actually we have

THEOREM 2.1.1. *Problem* (1.1) *has a unique solution. The solution is non-negative.*

Proof. Suppose p is a solution of (1.1). Using the definition of a solution to (1.1) we can obtain (by integration along the characteristic lines)

$$\begin{aligned} p(a,t) &= \exp\Big\{-\int_0^a \mu(s,t-a+s)ds\Big\} b(t-a) \\ &+ \int_0^a \exp\Big\{-\int_s^a \mu(\tau,t-a+\tau)d\tau\Big\} f(s,t-a+s)ds, \end{aligned} \tag{1.3}$$

for almost all $(a,t) \in Q_T$, $a < t$ and

$$\begin{aligned} p(a,t) &= \exp\Big\{-\int_0^t \mu(a-t+s,s)ds\Big\}p_0(a-t) \\ &+\int_0^t \exp\Big\{-\int_s^t \mu(a-t+\tau,\tau)d\tau\Big\}f(a-t+s,s)ds, \end{aligned} \tag{1.4}$$

for almost all $(a,t) \in Q_T$, $a > t$.

Here

$$b(t) = \int_0^{a_\dagger} \beta(a,t)p(a,t)da \quad \text{a.e. } t \in (0,T) \tag{1.5}$$

(see figure 2).

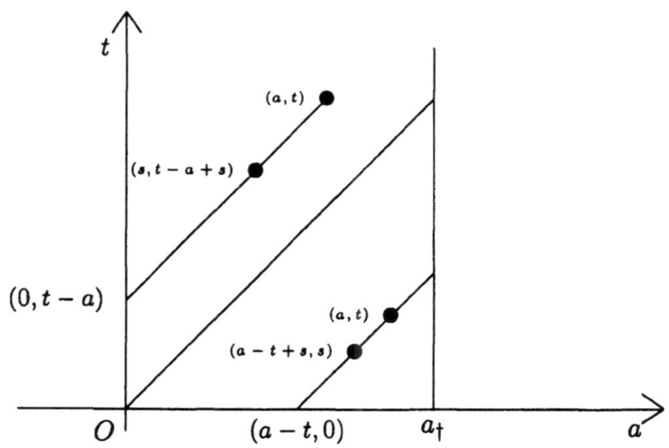

Figure 2

If we now assume $b \in L^\infty(0,T)$ as given, then p defined by (1.3) and (1.4) is the solution of

$$\begin{cases} Dp(a,t) + \mu(a,t)p(a,t) = f(a,t) & (a,t) \in Q_T, \\ p(0,t) = b(t) & t \in (0,T), \\ p(a,0) = p_0(a) & a \in (0,a_\dagger), \end{cases} \tag{1.6}$$

where a solution to (1.6) can be defined in the same manner as a solution to (1.1); that means

p is *a solution* of (1.6) if $p \in L^\infty(0,T;L^1(0,a_\dagger))$ is absolutely continuous along almost every characteristic line and satisfies

$$\begin{cases} Dp(a,t)+\mu(a,t)p(a,t)=f(a,t) & \text{a.e. in } Q_T, \\ \lim\limits_{\varepsilon\to 0^+} p(\varepsilon,t+\varepsilon)=b(t) & \text{a.e. } t\in(0,T), \\ \lim\limits_{\varepsilon\to 0^+} p(a+\varepsilon,\varepsilon)=p_0(a) & \text{a.e. } a\in(0,a_\dagger). \end{cases}$$

Since the solution p of (1.1) is given by (1.3)-(1.4), where b satisfies (1.5), we conclude that for almost all $0<t<\min\{T,a_\dagger\}$

$$\begin{aligned} b(t) &= \int_0^{a_\dagger} \beta(a,t)p(a,t)da \\ &= \int_0^t \beta(a,t)\exp\Big\{-\int_0^a \mu(s,t-a+s)ds\Big\}b(t-a)da \\ &+ \int_0^t \beta(a,t)\int_0^a \exp\Big\{-\int_s^a \mu(\tau,t-a+\tau)d\tau\Big\}f(s,t-a+s)ds\,da \\ &+ \int_t^{a_\dagger} \beta(a,t)\exp\Big\{-\int_0^t \mu(a-t+s,s)ds\Big\}p_0(a-t)da \\ &+ \int_t^{a_\dagger} \beta(a,t)\int_0^t \exp\Big\{-\int_s^t \mu(a-t+\tau,\tau)d\tau\Big\}f(a-t+s,s)ds\,da, \end{aligned}$$

and for almost all $\min\{T,a_\dagger\}<t<T$ (if there exists such a t; this happens only if $a_\dagger<T$)

$$\begin{aligned} b(t) &= \int_0^{a_\dagger} \beta(a,t)p(a,t)da \\ &= \int_0^{a_\dagger} \beta(a,t)\exp\Big\{-\int_0^a \mu(s,t-a+s)ds\Big\}b(t-a)da \\ &+ \int_0^{a_\dagger} \beta(a,t)\int_0^a \exp\Big\{-\int_s^a \mu(\tau,t-a+\tau)d\tau\Big\}f(s,t-a+s)ds\,da\,. \end{aligned}$$

Thus b satisfies the following Volterra equation

$$b(t)=F(t)+\int_0^t K(t,a)b(t-a)da \quad \text{a.e. } t\in(0,T),$$

or equivalently

$$b(t) = F(t) + \int_0^t K(t, t-s)b(s)ds \quad \text{a.e. } t \in (0, T), \tag{1.7}$$

with

$$\begin{aligned} F(t) &= \int_0^t \beta(a,t) \int_0^a \exp\{-\int_s^a \mu(\tau, t-a+\tau)d\tau\} f(s, t-a+s)ds\, da \\ &+ \int_t^{a_\dagger} \beta(a,t) \int_0^t \exp\{-\int_s^t \mu(a-t+\tau, \tau)d\tau\} f(a-t+s, s)ds\, da \\ &+ \int_t^{a_\dagger} \beta(a,t) \exp\{-\int_0^t \mu(a-t+s, s)ds\} p_0(a-t)da \end{aligned} \tag{1.8}$$

a.e. $t \in (0, \min\{T, a_\dagger\})$,

$$F(t) = \int_0^{a_\dagger} \beta(a,t) \int_0^a \exp\{-\int_s^a \mu(\tau, t-a+\tau)d\tau\} f(s, t-a+s)ds\, da \tag{1.9}$$

a.e. $\min\{T, a_\dagger\} < t < T$ (if there is such a t) and

$$K(t,a) = \begin{cases} \beta(a,t)e^{-\int_0^a \mu(s,t-a+s)ds} & \text{a.e. } (a,t) \in Q_T,\ a < t, \\ 0 & \text{elsewhere}. \end{cases} \tag{1.10}$$

Equation (1.7) is known as the *renewal equation* and also as the Lotka equation. From (1.8)-(1.10) and **(A1)-(A3)** it follows that

$$K \in L^\infty(Q_T), \quad K(t,a) \geq 0 \quad \text{a.e. } (a,t) \in Q_T \tag{1.11}$$

and

$$F \in L^\infty(0,T), \quad F(t) \geq 0 \quad \text{a.e. } t \in (0,T). \tag{1.12}$$

We shall prove via the Banach fixed point theorem that equation (1.7) has a unique solution $b \in L^\infty(0,T)$. Indeed, consider on $L^\infty(0,T)$ the norm

$$\|q\| = Ess\ sup_{t \in (0,T)}(e^{-\lambda t}|q(t)|), \quad \text{for any } q \in L^\infty(0,T),$$

which is equivalent to the usual norm on $L^\infty(0,T)$ (the constant $\lambda > 0$ will be made precise later) and the operator

$$\mathcal{F} : L^\infty(0,T) \to L^\infty(0,T)$$

given by

$$(\mathcal{F}q)(t) = F(t) + \int_0^t K(t, t-s)q(s)ds \quad \text{a.e. } t \in (0,T).$$

Now, for any $b_1,\ b_2 \in L^\infty(0,T)$ we have

$$\|(\mathcal{F}b_1)(t) - (\mathcal{F}b_2)(t)\|$$

$$= Ess\ sup_{t\in(0,T)} \left(e^{-\lambda t}\left|\int_0^t K(t,t-s)(b_1-b_2)(s)ds\right|\right)$$

$$\leq Ess\ sup_{t\in(0,T)} \left(e^{-\lambda t}\|K\|_{L^\infty(Q_T)} \int_0^t e^{\lambda s}e^{-\lambda s}|(b_1-b_2)(s)|ds\right)$$

$$\leq Ess\ sup_{t\in(0,T)} \left(e^{-\lambda t}\|K\|_{L^\infty(Q_T)} \cdot \|b_1-b_2\| \cdot \frac{1}{\lambda}e^{\lambda t}\right) .$$

It is obvious now that for any $\lambda > \|K\|_{L^\infty(Q_T)}$, $\mathcal{F}$ is a contraction on $(L^\infty(0,T), \|\cdot\|)$ and this implies the existence of a unique fixed point for $\mathcal{F}$, or equivalently of a unique solution $b \in L^\infty(0,T)$ for (1.7).

The Banach fixed point theorem also shows that the solution b of (1.7) is obtained via the iterative procedure

$$\begin{cases} b_0(t) = F(t) & \text{a.e. } t \in (0,T), \\ b_{n+1}(t) = F(t) + \displaystyle\int_0^t K(t,t-s)b_n(s)ds & \text{a.e. } t \in (0,T) . \end{cases}$$

The sequence $\{b_n\}$ is convergent in $(L^\infty(0,T), \|\cdot\|)$ to the solution of (1.7). Since by (1.11) and (1.12) we have $b_n(t) \geq 0$ a.e. $t \in (0,T)$, we may conclude that b, the solution of (1.7), is nonnegative on $(0,T)$.

Consider now p given by (1.3) and (1.4), where b is the unique solution of (1.7). Since $b(t) \geq 0$ a.e. in $(0,T)$, and using the assumptions **(A1)-(A3)**, we may conclude that $p(a,t) \geq 0$ a.e. in Q_T. As a consequence of (1.3) and (1.4) it immediately follows that $\mu p \in L^1(Q_T)$.

Indeed, let γ be an arbitrary characteristic line

$$\gamma = \{(a,t) \in \overline{Q}_T;\ a - t = k\}, \quad k \in \mathbf{R}.$$

Suppose that $k > 0$. Then

$$\gamma = \{(s+k,s) \in \mathbf{R}^2;\ s \in [0,\alpha]\},$$

where $(\alpha + k, \alpha) \in \{a_\dagger\} \times [0, T] \cup [0, a_\dagger] \times \{T\}$. So,

$$\int_\gamma \mu(a,t)p(a,t)dl = \int_0^\alpha \mu(s+k,s)e^{-\int_0^s \mu(\tau+k,\tau)d\tau} p_0(k)ds$$

$$+\int_0^\alpha \mu(s+k,s)\int_0^s e^{-\int_\tau^s \mu(\theta+k,\theta)d\theta} f(\tau+k,\tau)d\tau\ ds$$

$$\leq p_0(k) + \int_0^\alpha f(\tau+k,\tau)d\tau = p_0(k) + \int_\gamma f(a,t)dl.$$

If $k < 0$, then we obtain in the same manner that

$$\int_\gamma \mu(a,t)p(a,t)dl \leq b(-k) + \int_\gamma f(a,t)dl,$$

and in conclusion $\mu p \in L^1(Q_T)$ and

$$\|\mu p\|_{L^1(Q_T)} \leq C(\|p_0\|_{L^1(0,a_\dagger)} + \|b\|_{L^1(0,T)} + \|f\|_{L^1(Q_T)}),$$

where $C > 0$ is a constant.

From (1.3) and (1.4) it follows that p satisfies (1.2) and so we may infer that p is a solution of (1.1). The solution is unique because b, the solution of (1.7), is unique and is nonnegative because b, p_0, f are nonnegative.

REMARK 2.1.1. *If the mortality rate satisfies in addition*

$$(\mathbf{A4}) \qquad \begin{cases} \displaystyle\int_0^{a_\dagger} \mu(a, t - a_\dagger + a)da = +\infty \quad \text{a.e. } t \in (0,T), \\ \text{where } \mu(a,t) \text{ is extended by zero on } (0,a_\dagger) \times (-\infty, 0), \end{cases}$$

then the solution p of (1.1) is zero, for $a := a_\dagger$, i.e.,

$$\lim_{\varepsilon \to 0+} p(a_\dagger - \varepsilon, t - \varepsilon) = 0 \quad \text{a.e. } t \in (0,T)\ . \tag{1.13}$$

Note that **(A4)** is satisfied by all biological populations.

Indeed, by (1.4) we get that for almost any $0 < t < \min\{T, a_\dagger\}$ and for any $\varepsilon > 0$ small enough

$$p(a_\dagger - \varepsilon, t - \varepsilon) = \exp\Big\{-\int_0^{t-\varepsilon} \mu(a_\dagger - t + s, s)ds\Big\}p_0(a_\dagger - t)$$

$$+\int_0^{t-\varepsilon} \exp\Big\{-\int_s^{t-\varepsilon} \mu(a_\dagger - t + \tau, \tau)d\tau\Big\}f(a_\dagger - t + s, s)ds$$

$$\overset{\varepsilon\to 0^+}{\to} \exp\Big\{-\int_0^{t} \mu(a_\dagger - t + s, s)ds\Big\}p_0(a_\dagger - t)$$

$$+\int_0^{t} \exp\Big\{-\int_s^{t} \mu(a_\dagger - t + \tau, \tau)d\tau\Big\}f(a_\dagger - t + s, s)ds = 0,$$

(we have used the Lebesgue theorem) because ((**A4**) implies)

$$\int_0^t \mu(a_\dagger - t + s, s)ds = \int_{a_\dagger - t}^{a_\dagger} \mu(\tau, \tau + t - a_\dagger)d\tau$$

$$= \int_0^{a_\dagger} \mu(a, a + t - a_\dagger)da = +\infty \quad \text{a.e. } t \in (0, a_\dagger)$$

and

$$\int_s^t \mu(a_\dagger - t + \tau, \tau)d\tau = \int_{a_\dagger - t + s}^{a_\dagger} \mu(\tau, \tau + t - a_\dagger)d\tau$$

$$= \int_0^{a_\dagger} \mu(\tau, \tau + t - a_\dagger)d\tau - \int_0^{a_\dagger - t + s} \mu(\tau, \tau + t - a_\dagger)d\tau = +\infty$$

a.e. $(s, t) \in (0, a_\dagger) \times (0, \min\{T, a_\dagger\})$, $0 < s < t$. By (1.3) and (**A4**) we deduce that for almost all $\min\{T, a_\dagger\} < t < T$ (if there is such a t) and for any $\varepsilon > 0$ small enough

$$p(a_\dagger - \varepsilon, t - \varepsilon) = \exp\Big\{-\int_0^{a_\dagger - \varepsilon} \mu(s, t - a_\dagger + s)ds\Big\}b(t - a_\dagger)$$

$$+\int_0^{a_\dagger - \varepsilon} \exp\Big\{-\int_s^{a_\dagger - \varepsilon} \mu(\tau, t - \varepsilon + \tau)d\tau\Big\}f(s, t - a_\dagger + s)ds$$

$$\overset{\varepsilon\to 0^+}{\to} \exp\Big\{-\int_0^{a_\dagger} \mu(s, t - a_\dagger + s)ds\Big\}b(t - a_\dagger)$$

$$+\int_0^{a_\dagger} \exp\Big\{-\int_s^{a_\dagger} \mu(\tau, t - a_\dagger + \tau)d\tau\Big\}f(s, t - a_\dagger + s)ds = 0\,.$$

REMARK 2.1.2. *If in addition* $p_0 \in L^2(0,a_\dagger)$ *and* $f \in L^2(Q_T)$, *then from* (1.3) *and* (1.4) *it follows that* p, *the solution of* (1.1) *belongs to* $L^\infty(0,T;L^2(0,a_\dagger))$.

Moreover, if $p_0 \in L^\infty(0,a_\dagger)$ *and* $f \in L^\infty(Q_T)$, *then from* (1.3) *and* (1.4) *it follows that* $p \in L^\infty(Q_T)$.

We recall the classical Gronwall's lemma which will often be used throughout this book:

LEMMA (Gronwall). *Let* $x:[a,b]\to \mathbf{R}$ $(a,b\in\mathbf{R},\ a<b)$ *be a continuous function,* $\varphi \in L^\infty(a,b)$ *and* $\psi \in L^1(a,b)$, $\psi(t)\geq 0$ *a.e.* $t\in(a,b)$. *If*

$$x(t) \leq \varphi(t) + \int_a^t \psi(s)x(s)ds,$$

for each $t\in[a,b]$, *then*

$$x(t) \leq \varphi(t) + \int_a^t \varphi(s)\psi(s)\exp\left(\int_s^t \psi(\tau)d\tau\right)ds,$$

for each $t\in[a,b]$.

Proof. Denote by

$$y(t) = \int_a^t \psi(s)x(s)ds, \qquad t\in[a,b].$$

Using the equality

$$y'(t) = \psi(t)x(t) \qquad \text{a.e. in } (a,b),$$

and the hypotheses, it follows that

$$y'(t) \leq \psi(t)\varphi(t) + \psi(t)y(t) \qquad \text{a.e. in } (a,b).$$

Multiplying the last inequality by $\exp\left(-\int_a^t \psi(s)ds\right)$ we obtain that

$$\frac{d}{dt}\left[y(t)\exp\left(-\int_a^t \psi(s)ds\right)\right] \leq \psi(t)\varphi(t)\exp\left(-\int_a^t \psi(s)ds\right),$$

a.e. $t\in(a,b)$. By integration we conclude that

$$y(t) \leq \int_a^t \psi(s)\varphi(s)\left[\exp\left(\int_s^t \psi(\tau)d\tau\right)\right]ds,$$

for any $t\in[a,b]$ and consequently we get the conclusion of the lemma.

If in addition, $\varphi(t) = M$ for each $t \in [a,b]$, from the inequality above we deduce the following

LEMMA (Bellman). *If $x \in C([a,b])$, $\psi \in L^1(a,b)$, $\psi(t) \geq 0$ a.e. $t \in (a,b)$, $M \in \mathbf{R}$ and*

$$x(t) \leq M + \int_a^t \psi(s)x(s)ds$$

for each $t \in [a,b]$, then

$$x(t) \leq M \exp\left(\int_a^t \psi(s)ds\right)$$

for each $t \in [a,b]$.

THEOREM 2.1.2. *Let p be the solution of* (1.1). *We get*

(*i*) *if $f(a,t) > 0$ a.e. in Q_T, then $p(a,t) > 0$ a.e. in Q_T;*

(*ii*) *if β_i, μ_i, p_{0i}, f_i satisfy* **(A1)-(A3)** $(i = \overline{1,2})$ *and*

$$\beta_1(a,t) \geq \beta_2(a,t) \quad , \quad \mu_1(a,t) \leq \mu_2(a,t) \qquad \text{a.e. in } Q_T,$$

$$p_{01}(a) \geq p_{02}(a) \qquad \text{a.e. in } (0,a_\dagger),$$

$$f_1(a,t) \geq f_2(a,t) \qquad \text{a.e. in } Q_T,$$

then $p^1(a,t) \geq p^2(a,t)$ a.e. in Q_T, where p^i is the solution of (1.1), *corresponding to $\beta := \beta_i$, $\mu := \mu_i$, $p_0 := p_{0i}$, $f := f_i$, $i = \overline{1,2}$;*

(*iii*) *if $f_n \to f$ in $L^1(Q_T)$ (f_n satisfies* **(A3)**), *then*

$$p_n \to p$$

in $L^\infty(0,T;L^1(0,a_\dagger))$, where p_n are the solutions of (1.1) *corresponding to $f := f_n$, respectively.*

Proof. By Theorem 2.1.1, we get that the solution of (1.1) is given by (1.3) and (1.4), where b, the solution of (1.7), is nonnegative.

If $f(a,t) > 0$ a.e. in Q_T, then (1.3) and (1.4) imply that $p(a,t) > 0$ a.e. in Q_T.

The solution b of (1.7) can be obtained as the limit in $L^\infty(0,T)$ of the iterative sequence

$$\begin{cases} b_0^{\beta,\mu,p_0,f}(t) = F^{\beta,\mu,p_0,f}(t), \\ \\ b_{n+1}^{\beta,\mu,p_0,f}(t) = F^{\beta,\mu,p_0,f}(t) + \displaystyle\int_0^t K^{\beta,\mu,p_0,f}(t,t-s)b_n^{\beta,\mu,p_0,f}(s)ds \end{cases} \tag{1.14}$$

a.e. $t \in (0,T)$, $n \in \mathbf{N}^*$, where $F^{\beta,\mu,p_0,f}$ is given by (1.8) and (1.9) and $K^{\beta,\mu,p_0,f}$ is given by (1.10) (they both depend on β, μ, p_0, f).

As concerns (ii), if

$$\beta_1(a,t) \geq \beta_2(a,t) \quad , \quad \mu_1(a,t) \leq \mu_2(a,t) \qquad \text{a.e. in } Q_T,$$

$$p_{01}(a) \geq p_{02}(a) \qquad \text{a.e. in } (0,a_\dagger),$$

$$f_1(a,t) \geq f_2(a,t) \qquad \text{a.e. in } Q_T,$$

then

$$K^{\beta_1,\mu_1,p_{01},f_1}(t,a) \geq K^{\beta_2,\mu_2,p_{02},f_2}(t,a) \quad \text{a.e. in } Q_T,$$

$$F^{\beta_1,\mu_1,p_{01},f_1}(t) \geq F^{\beta_2,\mu_2,p_{02},f_2}(t) \quad \text{a.e. in } (0,T)$$

and by (1.14) we deduce $b^{\beta_2,\mu_2,p_{02},f_2}(t) \leq b^{\beta_1,\mu_1,p_{01},f_1}(t)$ a.e. in $(0,T)$. Now, using again (1.3) and (1.4) we may conclude that (ii) holds.

Finally, if $f_n \to f$ in $L^1(Q_T)$, then we infer that $F_n \to F$ in $L^\infty(0,T)$ (where F_n is given by (1.8)-(1.9), with $f := f_n$). Indeed, by (1.8)-(1.9) we get

$$F_n(t) - F(t)$$

$$= \int_0^t \beta(a,t) \int_0^a e^{-\int_s^a \mu(\tau,t-a+\tau)d\tau} (f_n - f)(s,t-a+s)ds\, da$$

$$+ \int_t^{a_\dagger} \beta(a,t) \int_0^t e^{-\int_s^t \mu(a-t+\tau,\tau)d\tau} (f_n - f)(a-t+s,s)ds\, da$$

a.e. $t \in (0, \min\{T,a_\dagger\})$, and

$$F_n(t) - F(t) = \int_0^{a_\dagger} \beta(a,t) \int_0^a e^{-\int_s^a \mu(\tau,t-a+\tau)d\tau} (f_n - f)(s,t-a+s)ds\, da$$

a.e. $\min\{T,a_\dagger\} < t < T$. Consequently we get

$$\|F_n - F\|_{L^\infty(0,T)} \leq \|\beta\|_{L^\infty(Q_T)} \cdot \|f_n - f\|_{L^1(Q_T)},$$

which implies that $F_n \to F$ in $L^\infty(0,T)$ as $n \to +\infty$.

Using (1.7) we may infer via Bellman's lemma that $b_{f_n} \to b$ in $L^\infty(0,T)$. Here b_{f_n} is the solution of (1.7) corresponding to $F := F_n$ (and consequently to $f := f_n$).

Using again (1.3) and (1.4) it follows that $p_n \to p$ in $L^\infty(0,T;L^1(0,a_\dagger))$ and thus (iii) is proved.

REMARK 2.1.3. *For any $f \in L^1(Q_T)$, $f(a,t) > 0$ a.e. in Q_T, the solution p of (1.1) satisfies*

$$p(a,t) > 0 \quad a.e.\ (a,t) \in Q_T$$

and (1.13) (by Remark 2.1.1).

These show that the biological meaning of $a_\dagger$ is the maximal age of the population.

By (1.3) and (1.4) we immediately conclude that if (1.13) is satisfied then **(A4)** *holds. So,* **(A4)** *is the necessary and sufficient condition to have $a_\dagger$ as the maximal age of the population species.*

Suppose now that the vital rates are time independent and that there is no inflow ($f \equiv 0$). Assumptions **(A1)**, **(A2)** and **(A4)** become

(A'1) $\beta \in L^\infty(0, a_\dagger),\ \beta(a) \geq 0$ a.e. $a \in (0, a_\dagger)$,

(A'2) $\mu \in L^1_{loc}([0, a_\dagger)),\ \mu(a) \geq 0$ a.e. $a \in (0, a_\dagger)$,

(A'4) $\displaystyle\int_0^{a_\dagger} \mu(a)da = +\infty$.

Assume that the initial data p_0 satisfies **(A3)** and in addition

(A'3) $p_0 \in W^{1,1}(0, a_\dagger),\ \mu p_0 \in L^1(0, a_\dagger)$,

(A"3) $\displaystyle p_0(0) = \int_0^{a_\dagger} \beta(a)p_0(a)da.$

Under these more restrictive assumptions the following result holds:

THEOREM 2.1.3. *The solution p of (1.1) belongs to $C([0, a_\dagger] \times [0, T])$ and the partial derivatives $\dfrac{\partial p}{\partial a}$ and $\dfrac{\partial p}{\partial t}$ exist almost everywhere in Q_T.*

Sketch of proof. By (1.8) and (1.9) we get

$$F(t) = \begin{cases} \displaystyle\int_t^{a_\dagger} \beta(a)e^{-\int_0^t \mu(a-t+s)ds} p_0(a-t)da, & t \in (0, \min\{T, a_\dagger\}), \\ 0, & \min\{T, a_\dagger\} < t < T \end{cases}$$

and after a simple verification we obtain that $F \in W^{1,\infty}(0,T)$ and (by (1.7)) that $b \in W^{1,\infty}(0,T)$ and

$$b'(t) = F'(t) + K(t)b(0) + \int_0^t K(t-s)b'(s)ds$$

a.e. $t \in (0,T)$. Here

$$K(a) = \begin{cases} \beta(a)\exp\{-\displaystyle\int_0^a \mu(s)ds\} & \text{a.e. } a \in (0, a_\dagger), \\ 0 & \text{elsewhere }. \end{cases}$$

Relations (1.3) and (1.4) imply now that $p \in C([0, a_\dagger] \times [0,T])$ and the partial derivatives $\dfrac{\partial p}{\partial a}$ and $\dfrac{\partial p}{\partial t}$ exist almost everywhere in Q_T.

Before ending this section, we note that, by (1.2) the solution $p \in L^\infty(0,T;L^1(0,a_\dagger))$ of problem (1.1) is also *a weak solution,* in the following sense:

$$\int_0^T \int_0^{a_\dagger} \{ - D\varphi(a,t) + \mu(a,t)\varphi(a,t) - \beta(a,t)\varphi(0,t)\}p(a,t)da\, dt$$

$$= \int_0^{a_\dagger} \varphi(a,0)p_0(a)da + \int_0^T \int_0^{a_\dagger} f(a,t)\varphi(a,t)da\, dt,$$

where φ is any absolutely continuous function along almost every characteristic line and satisfies

$$\begin{cases} \varphi \in L^\infty(Q_T), \\ D\varphi \in L^1(Q_T), \\ D\varphi - \mu\varphi + \beta\varphi(0,\cdot) \in L^\infty(Q_T), \\ \varphi(a_\dagger,t) = 0 \quad \text{a.e. } t \in (0,T)\ ,\ \varphi(a,T) = 0 \quad \text{a.e. } a \in (0,a_\dagger)\ . \end{cases} \tag{1.15}$$

Denote by Φ the set of all φ with the above mentioned properties.

Since φ is absolutely continuous along almost every characteristic line, it is obvious that the meanings of $\varphi(0,t)$ and $\varphi(a,0)$ are

$$\varphi(0,t) = \lim_{\varepsilon\to 0^+} \varphi(\varepsilon, t+\varepsilon) \quad \text{a.e. } t \in (0,T)$$

and

$$\varphi(a,0) = \lim_{\varepsilon\to 0^+} \varphi(a+\varepsilon, \varepsilon) \quad \text{a.e. } a \in (0,a_\dagger).$$

We can state now the following uniqueness result

THEOREM 2.1.4. *If* **(A1)-(A4)** *hold, then system* (1.1) *has a unique weak solution.*

Proof. Let p_1 and p_2 be two solutions of (1.1). Their difference $w = p_1 - p_2$ satisfies $w \in L^\infty(0,T;L^1(0,a_\dagger))$ and

$$\int_0^T \int_0^{a_\dagger} \big[- D\varphi(a,t) + \mu(a,t)\varphi(a,t) - \beta(a,t)\varphi(0,t)\big] w(a,t)da\, dt = 0, \tag{1.16}$$

for any $\varphi \in \Phi$.

We need to establish an auxiliary result:

LEMMA 2.1.5. *Let $g \in L^\infty(Q_T)$. Then the following problem*

$$\begin{cases} D\varphi - \mu(a,t)\varphi = g(a,t) - \beta(a,t)\varphi(0,t), & (a,t) \in Q_T, \\ \varphi(a_\dagger, t) = 0, & t \in (0,T), \\ \varphi(a,T) = 0, & a \in (0,a_\dagger) \end{cases} \tag{1.17}$$

has a unique solution φ.

If moreover $g(a,t) \geq 0$ a.e. in Q_T, then $\varphi(a,t) \leq 0$ a.e. in Q_T.

Here, a function φ, absolutely continuous along almost every characteristic line, is *a solution* to (1.17) if it satisfies $(1.15)_{1,2}$ and

$$\begin{cases} (D\varphi - \mu\varphi)(a,t) = g(a,t) - \beta(a,t)\varphi(0,t) & \text{a.e. } (a,t) \in Q_T, \\ \lim_{\varepsilon \to 0+} \varphi(a_\dagger - \varepsilon, t - \varepsilon) = 0 & \text{a.e. in } (0,T), \\ \lim_{\varepsilon \to 0+} \varphi(a - \varepsilon, T - \varepsilon) = 0 & \text{a.e. in } (0, a_\dagger)\ . \end{cases} \tag{1.18}$$

Proof of Lemma 2.1.5. If φ is a solution of (1.17), then for almost any $(a,t) \in Q_T$ such that $a - t > a_\dagger - T$ we have

$$\varphi(a,t) = \int_a^{a_\dagger} e^{-\int_a^s \mu(\tau, t-a+\tau)d\tau} \{\beta(s, t-a+s)B(t-a+s) - g(s, t-a+s)\} ds \tag{1.19}$$

and for almost any $(a,t) \in Q_T$ such that $a - t < a_\dagger - T$ we have

$$\varphi(a,t) = \int_t^T e^{-\int_t^s \mu(a-t+\tau,\tau)d\tau} \{\beta(a-t+s, s)B(s) - g(a-t+s, s)\} ds, \tag{1.20}$$

where $B \in L^\infty(0,T)$ is the solution of

$$B(t) = \begin{cases} \displaystyle\int_0^{a_\dagger} \exp\Big\{-\int_0^s \mu(\tau, t+\tau)d\tau\Big\}\beta(s,t+s)B(t+s)ds \\ \displaystyle\quad - \int_0^{a_\dagger} \exp\Big\{-\int_0^s \mu(\tau,t+\tau)d\tau\Big\} g(s,t+s)ds \quad \text{if } t < T - a_\dagger, \\ \displaystyle\int_t^T \exp\Big\{-\int_t^s \mu(\tau - t, \tau)d\tau\Big\}\beta(s-t,s)B(s)ds \\ \displaystyle\quad - \int_t^T \exp\Big\{-\int_t^s \mu(\tau - t,\tau)d\tau\Big\} g(s-t,s)ds \quad \text{if } t > T - a_\dagger. \end{cases}$$

It is obvious now that B is the solution of the following Volterra equation

$$B(t) = \int_t^T K(s, s-t)B(s)ds + G(t), \quad t \in (0,T), \tag{1.21}$$

where

$$G(t) = \begin{cases} -\displaystyle\int_0^{a_\dagger} \exp\{-\int_0^s \mu(\tau, t+\tau)d\tau\} g(s, t+s)ds & \text{if } t < T - a_\dagger, \\ -\displaystyle\int_t^T \exp\{-\int_t^s \mu(\tau - t, \tau)d\tau\} g(s-t, s)ds & \text{if } t > T - a_\dagger, \end{cases}$$

and K is defined by (1.10).

Using the Banach fixed point theorem we conclude the existence of a unique solution $B \in L^\infty(0,T)$ for (1.21), and it is obtained as the limit in $L^\infty(0,T)$ of the iterative sequence

$$\begin{cases} B_0(t) = G(t) & \text{a.e. } t \in (0,T), \\ B_{n+1}(t) = G(t) + \displaystyle\int_t^T K(s, s-t)B_n(s)ds & \text{a.e. } t \in (0,T), \end{cases} \tag{1.22}$$

$n \in \mathbf{N}^*$. The function φ defined by (1.19) and (1.20) (where B is the solution of (1.21)) belongs to $L^\infty(Q_T)$ and satisfies (1.18). After an easy calculation we conclude $\mu\varphi \in L^1(Q_T)$ and $\beta\varphi(0,\cdot) \in L^1(Q_T)$, and consequently the second condition in (1.15) is verified.

Now, if $g(a,t) \geq 0$ a.e in Q_T, then $G(t) \leq 0$ a.e. in $(0,T)$, and from (1.22) follows $B_n(t) \leq 0$ a.e. in $(0,T)$ and $B(t) \leq 0$ a.e. in $(0,T)$. Relations (1.19) and (1.20) allow us to conclude that $\varphi(a,t) \leq 0$ a.e. in Q_T.

Proof of Theorem 2.1.4 - continued. Consider the problem

$$\begin{cases} (D\varphi - \mu\varphi)(a,t) + \beta(a,t)\varphi(0,t) = sign\ w(a,t) & (a,t) \in Q_T, \\ \varphi(a,T) = 0 & a \in (0, a_\dagger), \\ \varphi(a_\dagger, t) = 0 & t \in (0,T). \end{cases} \tag{1.23}$$

Lemma 2.1.5 allows us to conclude that φ, the solution of (1.23) belongs to Φ. Putting now φ in (1.16) we get

$$\int_0^T \int_0^{a_\dagger} |w(a,t)| da\ dt = 0\ .$$

Thus, $w(a,t) = 0$ a.e. in Q_T and $p_1(a,t) = p_2(a,t)$ a.e. in Q_T.

2.2. A general nonlinear model

A more realistic description of the dynamics of a biological population is given by nonlinear age-dependent models. The birth and mortality rates are here nonlinear functions of the population size.

In this section we shall study some basic properties of the solution of the following nonlinear model:

$$\begin{cases} Dp(a,t) + \mu(a,t,\int_0^{a_\dagger} p(a,t)da)p(a,t) = f(a,t), & (a,t) \in Q_T, \\ p(0,t) = \displaystyle\int_0^{a_\dagger} \beta(a,t,\int_0^{a_\dagger} p(a,t)da)p(a,t)da, & t \in (0,T), \\ p(a,0) = p_0(a), & a \in (0,a_\dagger). \end{cases} \tag{2.1}$$

Here are the assumptions that we will make throughout this section:

(An1) $\forall s \in \mathbf{R}^+$ the functions $\beta(\cdot,\cdot,s)$ and $\mu(\cdot,\cdot,s)$ belong to $L^\infty(\overline{Q}_T)$ and $L^1_{loc}([0,a_\dagger) \times [0,T])$, respectively.
β and μ are locally Lipschitz functions with respect to the third variable, i.e., for any $M > 0$, there exists $L(M) > 0$ such that
$|\beta(a,t,s_1) - \beta(a,t,s_2)| \leq L(M) \cdot |s_1 - s_2|$,
$|\mu(a,t,s_1) - \mu(a,t,s_2)| \leq L(M) \cdot |s_1 - s_2|$, $\forall s_1, s_2 \in [0,M]$, a.e. $(a,t) \in Q_T$.

(An2) $\beta(a,t,x),\ \mu(a,t,x) \geq 0$ a.e. in $Q_T \times (0,+\infty)$.

(An3) $\beta(a,t,\cdot)$ is nonincreasing a.e. $(a,t) \in Q_T$ and
$\mu(a,t,\cdot)$ is nondecreasing a.e. $(a,t) \in Q_T$.

(An4) $p_0 \in L^1(0,a_\dagger),\ p_0(a) \geq 0$ a.e. $a \in (0,a_\dagger)$
$f \in L^1(Q_T),\ f(a,t) \geq 0$ a.e. $(a,t) \in Q_T$.

By *a solution* to (2.1) we mean a function $p \in L^\infty(0,T;L^1(0,a_\dagger))$, absolutely continuous along almost every characteristic line, which satisfies

$$\begin{cases} Dp(a,t) + \mu(a,t,P(t))p(a,t) = f(a,t) & \text{a.e. in } Q_T, \\ \lim\limits_{\varepsilon \to 0^+} p(\varepsilon, t+\varepsilon) = \displaystyle\int_0^{a_\dagger} \beta(a,t,P(t))p(a,t)da & \text{a.e. } t \in (0,T), \\ \lim\limits_{\varepsilon \to 0^+} p(a+\varepsilon,\varepsilon) = p_0(a) & \text{a.e. } a \in (0,a_\dagger), \end{cases}$$

where $P(t) = \displaystyle\int_0^{a_\dagger} p(a,t)da$ a.e. in $(0,T)$.

A solution to (2.1), for $T := +\infty$, is a function $p \in L^\infty_{loc}([0,+\infty);L^1(0,a_\dagger))$, which is a solution to (2.1) for any $T \in (0,+\infty)$.

Existence and uniqueness of a nonnegative solution to (2.1) will be proved via the Banach fixed point theorem.

First consider $P \in L^\infty(0,T)$, $P(t) \geq 0$ a.e. in $(0,T)$ arbitrary but fixed and let us study the system

$$\begin{cases} Dp(a,t) + \mu(a,t,P(t))p(a,t) = f(a,t), & (a,t) \in Q_T, \\ p(0,t) = \int_0^{a_\dagger} \beta(a,t,P(t))p(a,t)da, & t \in (0,T), \\ p(a,0) = p_0(a), & a \in (0,a_\dagger). \end{cases} \tag{2.2}$$

Problem (2.2) has a unique and nonnegative solution p (since the hypotheses on β, μ, p_0, f in Theorem 2.1.1 are fulfilled) and is given by

$$p(a,t;P) = \begin{cases} p_0(a-t)\Pi(a,t,t;P) \\ \qquad + \int_0^t f(a-s,t-s)\Pi(a,t,s;P)ds, \ a \geq t, \\ b(t-a;P)\Pi(a,t,a;P) \\ \qquad + \int_0^a f(a-s,t-s)\Pi(a,t,s;P)ds, \ a < t \end{cases} \tag{2.3}$$

(see (1.3) and (1.4) in the previous section), where

$$\Pi(a,t,s;P) = \exp\Big\{ - \int_0^s \mu(a-\tau,t-\tau,P(t-\tau))d\tau\Big\},$$

a.e. $s \in (0,\min\{a,t\})$ and $b(\cdot;P) \in L^\infty(0,T)$ is the solution of the Volterra integral equation

$$b(t;P) = F(t;P) + \int_0^t K(t,s;P)b(t-s;P)ds, \quad t \in (0,T). \tag{2.4}$$

Here we have set

$$K(t,a;P) = \beta(a,t,P(t))\Pi(a,t,a;P)$$

and

$$F(t;P) = \int_0^\infty \beta(a+t,t,P(t))p_0(a)\Pi(a+t,t,t;P)da$$

$$+ \int_0^\infty \beta(a,t,P(t)) \int_0^{\min\{a,t\}} f(a-s,t-s)\Pi(a,t,s;P)ds\ da,$$

where the functions p_0, β and Π are extended by zero outside their definition sets.

Note that the assumptions imply that $K \in L^\infty(Q_T \times (0,+\infty))$, $F \in L^\infty((0,T) \times (0,+\infty))$,

$$0 \leq K(t,a;P) \leq \beta(a,t,0) \text{ a.e. in } Q_T \times (0,+\infty),$$

$$0 \le F(t;P) \le F(t;0) \text{ a.e. in } (0,T) \times (0,+\infty).$$

We shall give the proof only for the case $T > a_\dagger$. We omit the proof for the case $T \le a_\dagger$ (which is analogous, but easier). Denote by

$$M_T = \max\{\|F(\cdot;0)\|_{L^\infty(0,T)} e^{T\|\beta(\cdot,\cdot;0)\|_{L^\infty(Q_T)}},\ \|p_0\|_{L^1(0,a_\dagger)}\} + \|f\|_{L^1(Q_T)}$$

and let

$$H = \{v \in L^\infty(0,T;L^1(0,a_\dagger));\ v(a,t) \ge 0 \text{ a.e. in } Q_T,$$

$$\|v(\cdot,t)\|_{L^1(0,a_\dagger)} \le M_T \text{ a.e. in } (0,T)\}$$

(which is a closed subset of $L^\infty(0,T;L^1(0,a_\dagger))$ and

$$\mathcal{H} = \{h \in L^\infty(0,T);\ 0 \le h(t) \le M_T \text{ a.e. in } (0,T)\}.$$

Before starting the proof of the existence and uniqueness of a solution to (2.1) we have to state some estimates.

LEMMA 2.2.1. *There exist M_{1T}, $M_{2T} > 0$ (constants depending on T) such that for any P_1, $P_2 \in \mathcal{H}$ we have*

$$|F(t;P_1) - F(t;P_2)| \le M_{1T}\left(|P_1(t) - P_2(t)| + \int_0^t |P_1(s) - P_2(s)|ds\right) \quad (2.5)$$

a.e. in $(0,T)$ and

$$\begin{cases} 0 \le b(t;P_1) \le M_{2T}, \\ |b(t;P_1) - b(t;P_2)| \le M_{2T}\left(|P_1(t) - P_2(t)| + \int_0^t |P_1(s) - P_2(s)|ds\right), \end{cases} \quad (2.6)$$

a.e. in $(0,T)$.

Proof. For any $P_1,\ P_2 \in \mathcal{H}$ we have that

$$|F(t;P_1) - F(t;P_2)|$$

$$\le \int_0^\infty |\beta(a+t,t,P_1(t)) - \beta(a+t,t,P_2(t))|p_0(a)\Pi(a+t,t,t;P_1)da$$

$$+\int_0^\infty \beta(a+t,t,P_2(t))p_0(a)|\Pi(a+t,t,t;P_1) - \Pi(a+t,t,t;P_2)|da$$

$$+\int_0^\infty |\beta(a,t,P_1(t)) - \beta(a,t,P_2(t))| \int_0^\gamma f(a-s,t-s)\Pi(a,t,s;P_1)ds\, da$$

$$+\int_0^\infty \beta(a,t,P_2(t)) \int_0^\gamma f(a-s,t-s)|\Pi(a,t,s;P_1) - \Pi(a,t,s;P_2)|ds\, da, \tag{2.7}$$

a.e. $t \in (0,T)$, where $\gamma = \min\{a,t\}$. Relation (2.7) implies that for almost any $t \in (a_\dagger, T)$ we get

$$|F(t;P_1) - F(t;P_2)| \le \int_0^{a_\dagger} L(M_T)|P_1(t) - P_2(t)| \int_0^a f(a-s,t-s)ds\, da$$

$$+\int_0^{a_\dagger} \beta(a,t,0) \int_0^a f(a-s,t-s)$$

$$\cdot \int_0^{a-s} |\mu(a-\tau,t-\tau,P_1(t-\tau)) - \mu(a-\tau,t-\tau,P_2(t-\tau))|d\tau\, ds\, da$$

$$\le L(M_T)|P_1(t) - P_2(t)| \cdot \|f\|_{L^1(Q_T)}$$

$$+L(M_T) \int_0^{a_\dagger} \beta(a,t,0) \int_0^a f(a-s,t-s)$$

$$\cdot \int_0^{a-s} |P_1(t-\tau) - P_2(t-\tau)|d\tau\, ds\, da$$

$$\le L(M_T)\|f\|_{L^1(Q_T)}|P_1(t) - P_2(t)|$$

$$+L(M_T)\|f\|_{L^1(Q_T)}\|\beta(\cdot,\cdot,0)\|_{L^\infty(Q_T)} \int_0^t |P_1(s) - P_2(s)|ds\, .$$

We conclude that for almost any $t \in (a_\dagger, T)$:

$$|F(t;P_1) - F(t;P_2)| \le L(M_T)\|f\|_{L^1(Q_T)} \max\{1, \|\beta(\cdot,\cdot,0)\|_{L^\infty(Q_T)}\}$$

$$\cdot(|P_1(t) - P_2(t)| + \int_0^t |P_1(s) - P_2(s)|ds). \tag{2.8}$$

By (2.7) we obtain in the same manner that for almost any $t \in (0, a_\dagger)$

$$|F(t;P_1) - F(t;P_2)| \le L(M_T)|P_1(t) - P_2(t)|\|p_0\|_{L^1(0,a_\dagger)}$$

$$+L(M_T)\|\beta(\cdot,\cdot,0)\|_{L^\infty(Q_T)}\|p_0\|_{L^1(0,a_\dagger)}\int_0^t |P_1(s) - P_2(s)|ds$$

$$+\int_0^{a_\dagger} |\beta(a,t,P_1(t)) - \beta(a,t,P_2(t))|$$

$$\cdot\int_0^\gamma f(a-s,t-s)\Pi(a,t,s;P_1)ds\ da$$

$$+\int_0^{a_\dagger} \beta(a,t,P_2(t))\int_0^\gamma f(a-s,t-s)$$

$$\cdot|\Pi(a,t,s;P_1) - \Pi(a,t,s;P_2)|ds\ da$$

$$\le L(M_T)\|p_0\|_{L^1(0,a_\dagger)}\max\{1, \|\beta(\cdot,\cdot,0)\|_{L^\infty(Q_T)}\}$$

$$\cdot\left(|P_1(t) - P_2(t)| + \int_0^t |P_1(s) - P_2(s)|ds\right)$$

$$+\|f\|_{L^1(Q_T)}L(M_T)|P_1(t) - P_2(t)|$$

$$+\|\beta(\cdot,\cdot,0)\|_{L^\infty(Q_T)}\|f\|_{L^1(Q_T)}L(M_T)\int_0^t |P_1(s) - P_2(s)|ds,$$

and so, for almost any $t \in (0, a_\dagger)$

$$|F(t;P_1) - F(t;P_2)| \le L(M_T)(\|p_0\|_{L^1(0,a_\dagger)} + \|f\|_{L^1(Q_T)})$$

$$\cdot\max\{1, \|\beta(\cdot,\cdot,0)\|_{L^\infty(Q_T)}\}\left(|P_1(t) - P_2(t)| + \int_0^t |P_1(s) - P_2(s)|ds\right). \tag{2.9}$$

Relations (2.8) and (2.9) imply that (2.5) holds, with

$$M_{1T} = L(M_T)(\|f\|_{L^1(Q_T)} + \|p_0\|_{L^1(0,a_\dagger)})\cdot\max\{1, \|\beta(\cdot,\cdot,0)\|_{L^\infty(Q_T)}\}.$$

By (2.4) and Gronwall's lemma we get

$$|b(t;P)| \le \|F(\cdot;0)\|_{L^\infty(0,T)}\exp\left\{T\|\beta(\cdot,\cdot;0)\|_{L^\infty(Q_T)}\right\} = \tilde{M}_{2T},$$

a.e. $t \in (0,T)$ and for any $P \in \mathcal{H}$.

Using again (2.4), we conclude that for almost all $t \in (0,T)$

$$
\begin{aligned}
|b(t;P_1) - b(t;P_2)| &\le |F(t;P_1) - F(t;P_2)| \\
&+ \int_0^{a_\dagger} |K(t,t-s;P_1) - K(t,t-s;P_2)| b(s;P_1)ds \\
&+ \int_0^{a_\dagger} K(t,t-s;P_2)|b(s;P_1) - b(s;P_2)|ds \\
&\le M_{1T}\left(|P_1(t) - P_2(t)| + \int_0^t |P_1(s) - P_2(s)|ds\right) \\
&+ \tilde{M}_{2T}(L(M_T)a_\dagger|P_1(t) - P_2(t)| \\
&+ L(M_T)a_\dagger\|\beta(\cdot,\cdot,0)\|_{L^\infty(Q_T)} \int_0^t |P_1(s) - P_2(s)|ds) \\
&+ \|\beta(\cdot,\cdot,0)\|_{L^\infty(Q_T)} \int_0^t |b(s;P_1) - b(s;P_2)|ds,
\end{aligned}
$$

and consequently

$$
\begin{aligned}
|b(t,P_1) - b(t,P_2)| \le M_{3T}&\left(|P_1(t) - P_2(t)| + \int_0^t |P_1(s) - P_2(s)|ds\right) \\
&+ \|\beta(\cdot,\cdot,0)\|_{L^\infty(Q_T)} \int_0^t |b(s;P_1) - b(s;P_2)|ds, \qquad (2.10)
\end{aligned}
$$

a.e. $t \in (0,T)$, where $M_{3T} = M_{1T} + \tilde{M}_{2T} L(M_T)a_\dagger \max\{1,\zeta\}$, $\zeta = \|\beta(\cdot,\cdot,0)\|_{L^\infty(Q_T)}$. From (2.10) and Gronwall's lemma follows (2.6), with

$$M_{2T} = \max\{\tilde{M}_{2T}, M_{3T}(2 + \zeta + T\zeta)e^{\zeta T}\}.$$

Let $\mathcal{T} : H \to L^\infty(0,T; L^1(0,a_\dagger))$, defined by

$$(\mathcal{T}q)(a,t) = p(a,t,Q),$$

where $Q(t) = \displaystyle\int_0^{a_\dagger} q(a,t)da$ a.e. $t \in (0,T)$. The following result for $\mathcal{T}$ holds:

LEMMA 2.2.2. *The function $\mathcal{T}$ maps H into itself and there exists $\overline{M}_T > 0$ (depending on T) such that*

$$\|(\mathcal{T}p_1)(t) - (\mathcal{T}p_2)(t)\|_{L^1(0,a_\dagger)} \le \overline{M}_T \int_0^t \|p_1(\cdot,s) - p_2(\cdot,s)\|_{L^1(0,a_\dagger)} ds, \quad (2.11)$$

a.e. $t \in (0,T)$, *for any* $p_1,\ p_2 \in H$.

Proof. For any $q \in H$, we take $Q(t) = \int_0^{a_\dagger} q(a,t)da$ a.e. $t \in (0,T)$. It is obvious that $Q \in L^\infty(0,T)$ and $Q(t) \geq 0$ a.e. $t \in (0,T)$. Let $b(\cdot;Q) \in L^\infty(0,T)$ be the solution of (2.4) on $(0,T)$ (which is nonnegative). Using Bellman's lemma we get

$$b(t;Q) \leq \|F(\cdot;0)\|_{L^\infty(0,T)} \exp\left\{\|K(\cdot,\cdot;0)\|_{L^\infty(Q_T)}T\right\} \tag{2.12}$$

a.e. $t \in (0,T)$. Using now (2.3) and (2.12) we get that $p(\cdot,\cdot,Q) \in H$ and so we conclude that $\mathcal{T}$ maps H into itself.

For any $p_1,\ p_2 \in H$ we define $P_i(t) = \int_0^{a_\dagger} p_i(a,t)da$ a.e. $t \in (0,T)$, $i = \overline{1,2}$, which obviously belong to $\mathcal{H}$. Using (2.3) and (2.4) we get that for almost any $t \in (0,a_\dagger)$

$$\|(\mathcal{T}p_1)(\cdot,t) - (\mathcal{T}p_2)(\cdot,t)\|_{L^1(0,a_\dagger)} = \|p(\cdot,t;P_1) - p(\cdot,t;P_2)\|_{L^1(0,a_\dagger)}$$

$$\leq \int_0^t |b(t-a;P_1) - b(t-a;P_2)|\Pi(a,t,a;P_1)da$$

$$+ \int_0^t |b(t-a;P_2)|\Pi(a,t,a;P_1) - \Pi(a,t,a;P_2)|da$$

$$+ \int_0^t \int_0^a f(a-s,t-s)|\Pi(a,t,s;P_1) - \Pi(a,t,s;P_2)|ds\ da$$

$$+ \int_t^{a_\dagger} p_0(a-t)|\Pi(a,t,t;P_1) - \Pi(a,t,t;P_2)|da$$

$$+ \int_t^{a_\dagger} \int_0^t f(a-s,t-s)|\Pi(a,t,s;P_1) - \Pi(a,t,s;P_2)|ds\ da.$$

Using now Lemma 2.2.1 and assumption **(An1)** we may infer that

$$\|(\mathcal{T}p_1)(\cdot,t) - (\mathcal{T}p_2)(\cdot,t)\|_{L^1(0,a_\dagger)}$$

$$\leq M_{2T} \int_0^t \left(|P_1(t-a) - P_2(t-a)| + \int_0^{t-a} |P_1(\tau) - P_2(\tau)|d\tau\right) da$$

$$+M_{2T}L(M_T)\int_0^t\int_0^a |P_1(t-s)-P_2(t-s)|ds\,da$$

$$+L(M_T)\int_0^t\int_0^a f(a-s,t-s)\int_0^s |P_1(t-\tau)-P_2(t-\tau)|d\tau\,ds\,da$$

$$+L(M_T)\int_t^{a_\dagger} p_0(a-t)\int_0^t |P_1(t-\tau)-P_2(t-\tau)|d\tau\,da$$

$$+L(M_T)\int_t^{a_\dagger}\int_0^t f(a-s,t-s)\int_0^s |P_1(t-\tau)-P_2(t-\tau)|d\tau\,ds\,da$$

$$\leq M_{2T}(1+T)\int_0^t |P_1(s)-P_2(s)|ds + M_{2T}L(M_T)T\int_0^t |P_1(s)-P_2(s)|ds$$

$$+L(M_T)\|f\|_{L^1(Q_T)}\int_0^t |P_1(s)-P_2(s)|ds$$

$$+L(M_T)\|p_0\|_{L^1(0,a_\dagger)}\int_0^t |P_1(s)-P_2(s)|ds$$

$$+L(M_T)\|f\|_{L^1(Q_T)}\int_0^t |P_1(s)-P_2(s)|ds$$

$$=\overline{M}_T\int_0^t |P_1(s)-P_2(s)|ds$$

$$\leq \overline{M}_T\int_0^t \|p_1(\cdot,s)-p_2(\cdot,s)\|_{L^1(0,a_\dagger)}ds,$$

where

$$\overline{M}_T = M_{2T}\left[1+T+TL(M_T)\right] + L(M_T)\|p_0\|_{L^1(0,a_\dagger)} + 2L(M_T)\|f\|_{L^1(Q_T)}.$$

In the same manner it is possible to prove that (2.11) holds, for almost any $t\in(a_\dagger,T)$ and for any $p_1,\ p_2\in H$.

We are now able to state the main result in this section:

THEOREM 2.2.3. *System* (2.1) *has a unique solution. This solution is non-negative.*

Proof. For any $\lambda > M_T$ arbitrary, but fixed, we can define the following equivalent norm on $L^\infty(0,T;L^1(0,a_\dagger))$:

$$\|q\| = Ess\ sup_{t\in(0,T)}\{e^{-\lambda t}\|q(\cdot,t)\|_{L^1(0,a_\dagger)}\}, \quad \forall q \in L^\infty(0,T;L^1(0,a_\dagger)).$$

We shall prove that $\mathcal{T}$ has a unique fixed point on H ((H,d) is a complete metric space, where $d(q_1,q_2) = \|q_1 - q_2\|$, $\forall q_1,\ q_2 \in H$). Indeed, for any $q_1,\ q_2 \in H$ we have (by Lemma 2.2.2)

$$\|\mathcal{T}q_1 - \mathcal{T}q_2\| = Ess\ sup_{(0,T)}\{e^{-\lambda t}\|(\mathcal{T}q_1)(\cdot,t) - (\mathcal{T}q_2)(\cdot,t)\|_{L^1(0,a_\dagger)}\}$$

$$\leq \overline{M}_T \cdot Ess\ sup_{t\in(0,T)}\{e^{-\lambda t}\int_0^t \|q_1(\cdot,s) - q_2(\cdot,s)\|_{L^1(0,a_\dagger)}ds\}$$
$$\leq \overline{M}_T Ess\ sup_{t\in(0,T)}\{e^{-\lambda t}\int_0^t e^{\lambda s}e^{-\lambda s}\|q_1(\cdot,s) - q_2(\cdot,s)\|_{L^1(0,a_\dagger)}ds\}$$

$$\leq \overline{M}_T\|q_1 - q_2\|\frac{1}{\lambda}\ .$$

It means that $\mathcal{T}$ is a contraction on (H,d) and consequently has a unique fixed point. It follows that (2.1) has a unique solution. By Theorem 2.1.1 we conclude that this solution is nonnegative.

REMARK 2.2.1. *If the mortality rate satisfies in addition*

$$\textbf{(An5)} \quad \begin{cases} \displaystyle\int_0^{a_\dagger} \mu(a, t - a_\dagger + a, 0)da = +\infty \quad \text{a.e. } t \in (0,T), \\ \text{where } \mu(a,t,0) \text{ is extended by zero on } (0,a_\dagger) \times (-\infty,0), \end{cases}$$

then the solution p of (2.1) *is zero, for $a := a_\dagger$, i.e.,*

$$\lim_{\varepsilon\to 0+} p(a_\dagger - \varepsilon, t - \varepsilon) = 0 \quad \text{a.e. } t \in (0,T).$$

Indeed, by (2.3) we have that for almost any $0 < t < a_\dagger$,

$$p(a_\dagger - \varepsilon, t - \varepsilon) = p_0(a_\dagger - t)e^{-\int_0^{t-\varepsilon}\mu(a_\dagger - t + s, s, P(s))ds}$$

$$+ \int_0^{t-\varepsilon} e^{-\int_s^{t-\varepsilon}\mu(a_\dagger - t+\tau,\tau,P(\tau))d\tau} f(a_\dagger - t + s, s)ds$$

$$\geq p_0(a_\dagger - t)e^{-\int_0^{t-\varepsilon} \mu(a_\dagger - t + s, s, 0)ds}$$

$$+\int_0^{t-\varepsilon} e^{-\int_s^{t-\varepsilon} \mu(a_\dagger - t + \tau, \tau, 0)d\tau} f(a_\dagger - t + s, s)ds \tag{2.13}$$

$$\overset{\varepsilon \to 0^+}{\to} 0,$$

where $P(t) = \int_0^{a_\dagger} p(a,t)da$ a.e. $t \in (0,T)$ (we have used the Lebesgue theorem and **(An5)**).

Now, for almost any $t \in (a_\dagger, T)$

$$p(a_\dagger - \varepsilon, t - \varepsilon) = b(t - a_\dagger)e^{-\int_0^{a_\dagger - \varepsilon} \mu(s, t - a_\dagger + s, P(t - a_\dagger + s))ds}$$

$$+\int_0^{a_\dagger - \varepsilon} e^{-\int_s^{a_\dagger - \varepsilon} \mu(\tau, t - a_\dagger + \tau, P(t - a_\dagger + \tau))d\tau} f(s, t - a_\dagger + s)ds$$

$$\geq b(t - a_\dagger)e^{-\int_0^{a_\dagger - \varepsilon} \mu(s, t - a_\dagger + s, 0)ds} \tag{2.14}$$

$$+\int_0^{a_\dagger - \varepsilon} e^{-\int_s^{a_\dagger - \varepsilon} \mu(\tau, t - a_\dagger + \tau, 0)d\tau} f(s, t - a_\dagger + s)ds$$

$$\overset{\varepsilon \to 0^+}{\to} 0$$

(by the same reason as in (2.13)). Using now (2.13) and (2.14) we get the conclusion.

An important particular situation is that of fertility rate β depending only on age a and time t and of the mortality rate μ with the structure

$$\mu(a, t, P) = \mu_0(a, t) + \mathcal{M}(P).$$

Here β and μ_0 are the natural fertility and mortality rates, respectively; $\mathcal{M}(P)$ is an additional mortality rate due to the limited resources of the habitat.

We wish to discuss the structure of the solution to the autonomous system

$$\begin{cases} Dp(a,t) + \mu_0(a,t)p(a,t) + \mathcal{M}(P(t))p(a,t) = 0 & (a,t) \in Q_T, \\ p(0,t) = \int_0^{a_\dagger} \beta(a,t)p(a,t)da & t \in (0,T), \\ p(a,0) = p_0(a) & a \in (0,a_\dagger), \\ P(t) = \int_0^{a_\dagger} p(a,t)da & t \in (0,T). \end{cases} \tag{2.15}$$

We shall use the following assumptions:

(**Ans1**) $\beta \in L^\infty(Q_T)$, $\beta(a,t) \geq 0$ a.e. in Q_T;

(**Ans2**)
$$\begin{cases} \mu_0 \in L^1_{loc}([0,a_\dagger) \times [0,T]) \ , \ \mu_0(a,t) \geq 0 \quad \text{a.e. in } Q_T \ , \\ \int_0^{a_\dagger} \mu_0(a, t - a_\dagger + a)da = +\infty \quad \text{a.e. } t \in (0,T), \\ \text{where } \mu_0(a,t) \text{ is extended by zero on } (0,a_\dagger) \times (-\infty, 0); \end{cases}$$

(**Ans3**) $\mathcal{M} : [0,+\infty) \to [0,+\infty)$ is continuously differentiable and nondecreasing function

and the initial density p_0 satisfies

(**Ans4**) $p_0 \in L^1(0,a_\dagger)$, $p_0(a) \geq 0$ a.e. in $(0,a_\dagger)$.

Problem (2.15) belongs to the class of "separable" models considered in [27]. By a similar procedure as in [10] we can get a solution of (2.15) under the separated form

$$p(a,t) = x(t)\tilde{p}(a,t),$$

where $\tilde{p}$ is the solution of the "free" problem

$$\begin{cases} D\tilde{p} + \mu_0(a,t)\tilde{p} = 0, & (a,t) \in Q_T, \\ \tilde{p}(0,t) = \int_0^{a_\dagger} \beta(a,t)\tilde{p}(a,t)da, & t \in (0,T), \\ \tilde{p}(a,0) = p_0(a), & a \in (0,a_\dagger), \end{cases}$$

which, we know, has a unique (and nonnegative) solution (see Section 2.1).

Denote by

$$P_0(t) = \int_0^{a_\dagger} \tilde{p}(a,t)da, \quad \text{a.e. } t \in (0,T).$$

It is obvious that $P_0 \in L^\infty(0,T)$ and $P_0(t) \geq 0$ a.e. in $(0,T)$.

Consider the following problem:

$$\begin{cases} x'(t) + \mathcal{M}(P_0(t)x(t))x(t) = 0, & t \in (0,T), \\ x(0) = x_0 \in (0,+\infty). \end{cases} \tag{2.16}$$

LEMMA 2.2.4. *System* (2.16) *has a unique Carathéodory solution.*

Proof. Denote by

$$C = \{v \in L^\infty(0,T);\ 0 \leq v(t) \leq x_0 \text{ a.e. } t \in (0,T)\}.$$

It is easy to prove via the Banach fixed point theorem that $\mathcal{T} : C \to C$ defined by $\mathcal{T}h = x_h$, where x_h is the Carathéodory solution of

$$\begin{cases} x'(t) + \mathcal{M}(P_0(t)h(t))x(t) = 0, & t \in (0,T), \\ x(0) = x_0, \end{cases}$$

has a unique fixed point. So, we may conclude that (2.16) has a unique Carathéodory solution.

THEOREM 2.2.5. *The function* $p \in L^\infty(0,T; L^1(0,a_\dagger))$ *given by*

$$p(a,t) = x(t)\tilde{p}(a,t), \quad \text{a.e. } (a,t) \in Q_T, \tag{2.17}$$

where x *is the Carathéodory solution of* (2.16) *with* $x_0 := 1$, *is the unique solution to* (2.15).

Proof. p given by (2.17) satisfies

$$p \in L^\infty(0,T; L^1(0,a_\dagger)), \quad p(a,t) \geq 0 \quad \text{a.e. in } Q_T$$

and is a solution of (2.15). The uniqueness of the solution to (2.15) allows us to conclude that the structure of this solution is given by (2.17), where x is the Carathéodory solution of (2.16), corresponding to $x_0 := 1$.

Note that Theorem 2.2.5 implies the nonnegativity of the solution to (2.15).

2.3. Asymptotic behaviour of the solution

Here we investigate the asymptotic behaviour of the solution of the following problem:

$$\begin{cases} Dp(a,t) + \mu(a)p(a,t) = 0, & (a,t) \in Q, \\ p(0,t) = \int_0^{a_\dagger} \beta(a)p(a,t)da, & t \in (0,+\infty), \\ p(a,0) = p_0(a), & a \in (0,a_\dagger), \end{cases} \tag{3.1}$$

where $Q = (0,a_\dagger) \times (0,+\infty)$. This is the model with time-independent vital rates and without inflow.

Assume that β, μ and p_0 satisfy

(A1) $\beta \in L^\infty(0,a_\dagger)$, $\beta(a) \geq 0$ a.e. $a \in (0,a_\dagger)$, β is not identically 0;

(A2) $\mu \in L^1_{loc}([0,a_\dagger))$, $\mu(a) \geq 0$ a.e. $a \in (0,a_\dagger)$,

$$\int_0^{a_\dagger} \mu(a)da = +\infty;$$

(A3) $p_0 \in L^1(0,a_\dagger)$, $p_0(a) \geq 0$ a.e. $a \in (0,a_\dagger)$.

Under these hypotheses there exists a unique solution p to (3.1), which is nonnegative (by Theorem 2.1.1). This solution satisfies

$$p(a,t) = b(t-a)\exp\{-\int_0^a \mu(s)ds\},$$

for almost any $(a,t) \in Q$, $a < t$, where b is the solution of the Volterra integral equation

$$b(t) = F(t) + \int_0^t K(t-s)\,b(s)\,ds \quad \text{a.e. } t \in (0,+\infty), \tag{3.2}$$

Here

$$F(t) = \begin{cases} \int_t^{a_\dagger} \beta(a)e^{-\int_0^t \mu(a-t+s)ds} p_0(a-t)da, & t \in (0,a_\dagger), \\ 0, & t \in (a_\dagger,+\infty) \end{cases} \tag{3.3}$$

and

$$K(t) = \begin{cases} \beta(t)\exp\{-\int_0^t \mu(s)ds\}, & t \in (0,a_\dagger), \\ 0, & \textit{elsewhere}. \end{cases} \tag{3.4}$$

It is obvious that $F \in C(\mathbf{R}^+)$, $K \in L^\infty(\mathbf{R}^+)$ and as a consequence $b \in C(\mathbf{R}^+)$. So, we have that p is a continuous function on $\{(a,t) \in Q;\ a < t\}$.

First we shall discuss the asymptotic behaviour of b. By (3.2)-(3.4) we may infer via Bellman's lemma that

$$|b(t)| \leq \|\beta\|_{L^\infty(0,a_\dagger)} \cdot \|p_0\|_{L^1(0,a_\dagger)} \cdot e^{t\|\beta\|_{L^\infty(0,a_\dagger)}}, \quad \forall t \in \mathbf{R}^+$$

and this implies that b is absolutely Laplace transformable.

By (3.2) we obtain that

$$\mathcal{L}(b)(\lambda) = \mathcal{L}(F)(\lambda) + \int_0^\infty e^{-\lambda t} \int_0^t K\,(t-s)\,b\,(s)\,ds\;dt,$$

where $\mathcal{L}(g)(\lambda)$ denotes the Laplace transform of g in $\lambda \in \mathbf{C}$. Namely,

$$\mathcal{L}(g)(\lambda) = \int_0^\infty e^{-\lambda t} g(t)dt.$$

Since F and K vanish for $t > a_\dagger$, their transforms $\mathcal{L}(F)$ and $\mathcal{L}(K)$ are entire analytical functions of $\lambda \in \mathbf{C}$.

The last relations imply

$$\begin{aligned}\mathcal{L}(b)(\lambda) &= \mathcal{L}(F)(\lambda) + \int_0^\infty \int_s^\infty e^{-\lambda t} K(t-s) b(s)dt\;ds \\ &= \mathcal{L}(F)(\lambda) + \mathcal{L}(b)(\lambda)\mathcal{L}(K)(\lambda),\end{aligned}$$

and in conclusion

$$\mathcal{L}(b)(\lambda) = \frac{\mathcal{L}(F)(\lambda)}{1-\mathcal{L}(K)(\lambda)} = \mathcal{L}(F)(\lambda) + \frac{\mathcal{L}(F)(\lambda)\mathcal{L}(K)(\lambda)}{1-\mathcal{L}(K)(\lambda)}. \tag{3.5}$$

We shall use classical Laplace transform techniques in order to study the asymptotic behaviour of b, which is related to the singularities of $\mathcal{L}(b)(\lambda)$.

By (3.5) we conclude that $\mathcal{L}(b)(\lambda)$ can have poles only among the roots of the equation

$$\mathcal{L}(K)(\lambda) = 1. \tag{3.6}$$

LEMMA 2.3.1. *Equation (3.6) has a unique real solution* α^**, which is a simple root. This solution is negative if and only if* $\int_0^{a_\dagger} K(s)ds < 1$*. Any other solution* α *of (3.6) satisfies* $\Re\alpha < \alpha^*$*.*

Proof. The function $\lambda \mapsto \mathcal{L}(K)(\lambda)$ is strictly decreasing on $\mathbf{R}$ because

$$\mathcal{L}(K)'(\alpha) = -\int_0^\infty te^{-\alpha t} K(t)dt < 0$$

for any $\alpha \in \mathbf{R}$ and satisfies

$$\lim_{\lambda \to -\infty} \mathcal{L}(K)(\lambda) = +\infty,$$

$$\lim_{\lambda \to \infty} \mathcal{L}(K)(\lambda) = 0.$$

It follows that there exists a unique $\alpha^* \in \mathbf{R}$ such that

$$\mathcal{L}(K)(\alpha^*) = \int_0^{a_\dagger} \beta(a) e^{-\alpha^* a - \int_0^a \mu(s)ds} da = 1$$

and since $\mathcal{L}(K)'(\alpha^*) < 0$ we get that α^* is simple. It is obvious that $\alpha^* < 0$ if and only if $\mathcal{L}(K)(0) = \int_0^\infty K(s)ds < 1$.

Let $\alpha = u_0 + iv_0$ ($u_0, v_0 \in \mathbf{R}$) be another solution of (3.6). Then

$$1 = \int_0^\infty e^{-\alpha^* t} K(t)\, dt = \int_0^\infty e^{-\alpha t} K(t)\, dt$$

$$= \int_0^\infty e^{-u_0 t} \cos(v_0 t)\, K(t)\, dt < \int_0^\infty e^{-u_0 t} K(t)\, dt$$

and as a consequence we get that $u_0 = \Re\alpha < \alpha^*$.

Now we are ready to formulate the result concerning the asymptotic behaviour of b:

THEOREM 2.3.2. *The solution b of (3.2) satisfies*

$$b(t) = e^{\alpha^* t} b_0(t), \quad \forall t \in \mathbf{R}^+, \tag{3.7}$$

where $\lim_{t \to +\infty} b_0(t) = u \geq 0$.

Proof. Since, by a classical result we get that

$$\lim_{|\lambda| \to +\infty, \Re\lambda > \alpha} \mathcal{L}(K)(\lambda) = \lim_{|\lambda| \to +\infty, \Re\lambda > \alpha} \mathcal{L}(F)(\lambda) = 0, \ \forall \alpha \in \mathbf{R}, \tag{3.8}$$

(see [34]) we conclude that

$$\lim_{|\lambda| \to +\infty, \Re\lambda > \alpha} \frac{\mathcal{L}(F)(\lambda)\mathcal{L}(K)(\lambda)}{1 - \mathcal{L}(K)(\lambda)} = 0.$$

On the other hand

$$\int_{-\infty}^\infty \left| \frac{\mathcal{L}(F)(x+iy)\mathcal{L}(K)(x+iy)}{1 - \mathcal{L}(K)(x+iy)} \right| dy < +\infty, \tag{3.9}$$

where $x \in \mathbf{R}$ is such that $\mathcal{L}(K)(x+iy) \neq 1$, $\forall y \in \mathbf{R}$. Indeed, relation (3.8) implies that

$$m(x) = \inf_{y \in \mathbf{R}} |1 - \mathcal{L}(K)(x+iy)| > 0.$$

Define the functions

$$f_x(t) = \begin{cases} e^{-xt} F(t), & t \geq 0, \\ 0, & t < 0, \end{cases}$$

and

$$k_x(t) = \begin{cases} e^{-xt} K(t), & t \geq 0, \\ 0, & t < 0. \end{cases}$$

Since f_x and k_x vanish outside of $[0, a_\dagger]$, their Fourier transforms $\widehat{f_x}$, $\widehat{k_x}$ belong to $L^2(\mathbf{R})$ and we can also verify that

$$\widehat{f_x}(y) = \mathcal{L}(F)(x+iy), \quad \widehat{k_x}(y) = \mathcal{L}(K)(x+iy).$$

Thus

$$\left| \frac{\mathcal{L}(F)(x+iy)\mathcal{L}(K)(x+iy)}{1 - \mathcal{L}(K)(x+iy)} \right| \leq \frac{1}{m(x)} \left| \widehat{f_x}(y) \widehat{k_x}(y) \right|$$

and so

$$\int_{-\infty}^{\infty} \left| \frac{\mathcal{L}(F)(x+iy)\mathcal{L}(K)(x+iy)}{1 - \mathcal{L}(K)(x+iy)} \right| dy \leq \frac{1}{m(x)} \left\| \widehat{f_x} \right\|_{L^2(\mathbf{R})} \cdot \left\| \widehat{k_x} \right\|_{L^2(\mathbf{R})},$$

which shows that (3.9) is satisfied.

For any (but fixed) $x > \alpha^*$ we consider the function

$$g(t) = \frac{1}{2\pi i} \int_{x-i\infty}^{x+i\infty} e^{\lambda t} \frac{\mathcal{L}(F)(\lambda)\mathcal{L}(K)(\lambda)}{1 - \mathcal{L}(K)(\lambda)} d\lambda, \tag{3.10}$$

which has the Laplace transform

$$\mathcal{L}(g)(\lambda) = \frac{\mathcal{L}(F)(\lambda)\mathcal{L}(K)(\lambda)}{1 - \mathcal{L}(K)(\lambda)}$$

(this follows by a classical result from Laplace transform theory; see [34]). Using now (3.5) we may infer that

$$\mathcal{L}(b)(\lambda) = \mathcal{L}(F)(\lambda) + \mathcal{L}(g)(\lambda)$$

and consequently

$$b(t) = F(t) + g(t), \quad t \in \mathbf{R}^+.$$

So, the asymptotic behaviour of b is given by the asymptotic behaviour of g.

Finally, consider $x_1 < \alpha^*$ such that any root λ of (3.6), other than α^*, satisfies $\Re\lambda < x_1$. By (3.8) and (3.9) we shift the integration in (3.10), from x to x_1. We get now that

$$g(t) = e^{\alpha^* t}(u + w(t)),$$

where

$$\begin{aligned} u &= Res\left(\frac{\mathcal{L}(F)\mathcal{L}(K)}{1-\mathcal{L}(K)}\right)(\alpha^*) \\ &= \mathcal{L}(F)(\alpha^*)\mathcal{L}(K)(\alpha^*)Res\left(\frac{1}{1-\mathcal{L}(K)}\right)(\alpha^*) \\ &= \frac{\int_0^\infty e^{-\alpha^* t}F(t)dt}{\int_0^\infty te^{-\alpha^* t}K(t)dt}, \end{aligned}$$

and

$$\begin{aligned} |w(t)| &= \frac{e^{-\alpha^* t}}{2\pi}\left|\int_{x_1-i\infty}^{x_1+i\infty}\frac{\mathcal{L}(F)(\lambda)\mathcal{L}(K)(\lambda)}{1-\mathcal{L}(K)(\lambda)}e^{\lambda t}d\lambda\right| \\ &\leq \frac{e^{-(\alpha^*-x_1)t}}{m(x_1)}\left\|\widehat{f}_{x_1}\right\|_{L^2(\mathbf{R})}\cdot\left\|\widehat{k}_{x_1}\right\|_{L^2(\mathbf{R})}. \end{aligned}$$

It is obvious that $u = 0$ if and only if $F(t) = 0$, $\forall t \geq 0$.

If $F(t) = 0$, $\forall t \geq 0$, then $b(t) = 0$, $\forall t \geq 0$, and so (3.7) is satisfied.

If $u \neq 0$, then we obtain that

$$b(t) = ue^{\alpha^* t}\left(1 + \frac{e^{-\alpha^* t}}{u} + \frac{w(t)}{u}\right), \quad \forall t \in \mathbf{R}^+$$

and as a consequence we get (3.7).

REMARK 2.3.1. *By (3.3) we get that $u = 0$ if and only if for all $t \geq 0$ we have*

$$\beta(a+t)p_0(a) = 0 \text{ a.e. } a \in (0, a_\dagger),$$

where β is extended by 0 outside $[0, a_\dagger]$.

This happens if the support of β lies to the left of the support of p_0 (p_0 is called in this case a trivial datum). In this case we get

$$p(a,t) = 0, \quad \forall (a,t) \in Q, \ a < t$$

and as a consequence

$$\lim_{t\to\infty}\|p(t)\|_{L^\infty(0,a_\dagger)} = 0.$$

Now we state the following theorem:

THEOREM 2.3.3.

$$p(a,t) = e^{\alpha^*(t-a)} b_0(t-a) \exp\left\{-\int_0^a \mu(s)\,ds\right\},$$

for almost any $(a,t) \in Q$, $a < t$.

REMARK 2.3.2. *In the case of a non-trivial datum* $(u > 0)$, *we may infer via Theorem 2.3.2 and Theorem 2.3.3 that*

1) $\lim_{t\to\infty} \|p(t)\|_{L^\infty(0,a_\dagger)} = 0$, if $\alpha^* < 0$,

2) $\lim_{t\to\infty} \|p(t)\|_{L^1(0,a_\dagger)} = +\infty$, if $\alpha^* > 0$,

3) $\lim_{t\to\infty} \|p(t) - \widetilde{p}\|_{L^\infty(0,a_\dagger)} = 0$, if $\alpha^* = 0$,

where

$$\widetilde{p}(a) = u \exp\left\{-\int_0^a \mu(s)\,ds\right\}, \quad \forall a \in (0, a_\dagger)$$

and is a nontrivial steady-state of (3.1).

Using now this remark and the analogy with the asymptotic behaviour of the solution to the malthusian model, we can call α^* the intrinsic malthusian parameter.

REMARK 2.3.3. *If we denote by*

$$R = \int_0^{a_\dagger} \beta(a)\, e^{-\int_0^a \mu(s)ds}\, da$$

(which is called the net reproduction rate), then the following relationship between the net reproduction rate and the intrinsic malthusian parameter holds:

$$\begin{cases} R > 1, & \text{if } \alpha^* > 0, \\ R = 1, & \text{if } \alpha^* = 0, \\ R < 1, & \text{if } \alpha^* < 0. \end{cases}$$

In the second part of this section we shall deal with the asymptotic behaviour of the solution of the following separable nonlinear model:

$$\begin{cases} Dp(a,t) + \mu(a)\,p(a,t) + \mathcal{M}(P(t))\,p(a,t) = 0, & (a,t) \in Q, \\ p(0,t) = \int_0^{a_\dagger} \beta(a)\,p(a,t)\,da, & t \in (0,+\infty), \\ p(a,0) = p_0(a), & a \in (0,a_\dagger), \end{cases} \tag{3.11}$$

where $P(t) = \int_0^{a_\dagger} p(a,t)\,da$, $t \in (0,+\infty)$ and

(A4) $\mathcal{M} : [0,+\infty) \to [0,+\infty)$ is a continuously differentiable and strictly increasing function,
$\mathcal{M}(0) = 0$ and $\lim_{r\to\infty} \mathcal{M}(r) = +\infty$.

By Theorem 2.2.5 we conclude that (3.11) has a unique solution p given by

$$p(a,t) = x(t)\,\widetilde{p}(a,t) \quad \text{a.e. } (a,t) \in Q,$$

where $\widetilde{p}$ is the solution of (3.1) and x is the Carathéodory solution to

$$\begin{cases} x'(t) + \mathcal{M}(P_0(t)x(t))\,x(t) = 0, & t \in (0,+\infty), \\ x(0) = 1, \end{cases}$$

where

$$P_0(t) = \int_0^{a_\dagger} p(a,t)\,da, \qquad t \in (0,+\infty).$$

It is obvious that x is a decreasing and nonnegative function. So, there exists $\lim_{t\to\infty} x(t) \in [0,1]$.

If p_0 is a trivial datum, then for any $t > a_\dagger$ we have

$$p(a,t) = 0, \quad \forall a \in (0,a_\dagger).$$

Otherwise we have

THEOREM 2.3.4. *If p_0 is a non-trivial datum, then*

$$\lim_{t\to+\infty} \|p(t) - \overline{p}\|_{L^\infty(0,a_\dagger)} = 0,$$

where $\overline{p}$ is the unique steady-state of (3.11).

Proof. If $\alpha^* < 0$, then $\overline{p} = 0$ is the unique steady-state of (3.1) and by Remark 2.3.2 we get the conclusion.

If $\alpha^* > 0$, then Theorem 2.3.3 implies that

$$P_0(t) = e^{\alpha^* t} h(t), \quad t \in (0,+\infty),$$

and $\lim_{t\to\infty} h(t) = h_0 > 0$. So, x is the solution of

$$\begin{cases} x'(t) + \mathcal{M}\left(e^{\alpha^* t} h(t)\,x(t)\right) x(t) = 0, & t \in (0,+\infty), \\ x(0) = 1, \end{cases}$$

or equivalently $y(t) = e^{\alpha^* t} x(t)$ $\left(t \in \mathbf{R}^+\right)$ is the solution of

$$\begin{cases} y'(t) = (\alpha^* - \mathcal{M}(h(t)y(t)))\, y(t), & t \in (0, +\infty), \\ y(0) = 1. \end{cases}$$

It is obvious that $y(t) > 0$, $\forall t \in [0, +\infty)$.

We shall prove that $\lim_{t\to\infty} y(t) = z_0$, where $z_0 \in (0, +\infty)$ is the unique solution of the equation

$$\alpha^* - \mathcal{M}(h_0 z_0) = 0.$$

For any $h > 0$ we denote by $z(h)$ the unique solution of the equation

$$\alpha^* - \mathcal{M}(hz) = 0.$$

Let us take any $0 < h_1 < h_0 < h_2$. Since $\mathcal{M}$ is strictly increasing we deduce that

$$z(h_1) > z_0 > z(h_2),$$

As $\lim_{t\to\infty} h(t) = h_0$, we infer that there exists $T \in (0, +\infty)$ such that

$$h_1 < h(t) < h_2, \ \forall t \in [T, +\infty).$$

We shall prove that

$$\lim_{t\to\infty} dist\,(y(t),\ [z(h_2), z(h_1)]) = 0.$$

Indeed, if for a certain $t \in [T, +\infty)$ we have

$$y(t) > z(h_1),$$

then

$$y'(t) = (\alpha^* - \mathcal{M}(h(t)\,y(t)))\,y(t) < (\alpha^* - \mathcal{M}(h_1 z(h_1)))\,y(t) = 0,$$

and so y is strictly decreasing on $[t, t+\varepsilon]$ $(\varepsilon > 0)$.

If for a certain $t \in [T, +\infty)$ we have

$$y(t) < z(h_2),$$

then

$$y'(t) > 0,$$

and so y is strictly increasing on $[t, t+\varepsilon]$ $(\varepsilon > 0)$. We conclude that

$$\lim_{t\to\infty} dist\,(y(t), [z(h_2), z(h_1)]) = \gamma \in [0, +\infty).$$

Since, $\lim_{h \to h_0} z(h) = z_0$, we obtain that

$$\lim_{t \to \infty} |y(t) - z_0| = \gamma_0 \in [0, +\infty),$$

and in conclusion there exists

$$\lim_{t \to \infty} y(t) = z_0 + \gamma_0 \text{ or } z_0 - \gamma_0.$$

If $\lim_{t \to \infty} y(t) = z_0 + \gamma_0$, then for t large enough we have

$$\exists \lim_{t \to \infty} y'(t) = (\alpha^* - \mathcal{M}(h_0(z_0 + \gamma_0)))(z_0 + \gamma_0).$$

The conclusion is that $\gamma_0 = 0$ and $\lim_{t \to \infty} y(t) = \gamma_0$.

If $\lim_{t \to \infty} y(t) = z_0 - \gamma_0$, then $\gamma_0 = 0$, otherwise, for t large enough we have $y'(t) > 0$ and so $z_0 - \gamma_0 > 0$. We conclude that

$$\exists \lim_{t \to \infty} y'(t) > 0,$$

which is absurd.

So, $\lim_{t \to \infty} y(t) = z_0$. It follows (by Theorem 2.3.3 and Remark 2.3.2) that

$$\lim_{t \to +\infty} \|p(t) - \overline{p}\|_{L^\infty(0,a_\dagger)} = 0,$$

where $\overline{p}(a) = uz_0 \exp\left\{-\alpha^* a - \int_0^a \mu(s)\,ds\right\}$, $a \in (0, a_\dagger)$. Here

$$h_0 = u \int_0^{a_\dagger} \exp\left\{-\alpha^* a - \int_0^a \mu(s)\,ds\right\} da$$

and

$$\overline{p}(a) = uz_0 \exp\left\{-\alpha^* a - \int_0^a \mu(s)\,ds\right\} = \frac{h_0 z_0 \exp\left\{-\alpha^* a - \int_0^a \mu(s)\,ds\right\}}{\int_0^{a_\dagger} \exp\left\{-\alpha^* a - \int_0^a \mu(s)\,ds\right\} da}$$

$$= \frac{\mathcal{M}^{-1}(\alpha^*) \exp\left\{-\alpha^* a - \int_0^a \mu(s)\,ds\right\}}{\int_0^{a_\dagger} \exp\left\{-\alpha^* a - \int_0^a \mu(s)\,ds\right\} da}, \quad \forall a \in [0, a_\dagger],$$

is the unique steady state of (3.1).

In the same manner follows the conclusion for $\alpha^* = 0$.

2.4. A linear periodic age-dependent population dynamics

This section is devoted to the basic properties of the solution of the following linear model describing a periodic age-dependent population dynamics:

$$\begin{cases} Dp(a,t) + \mu(a,t)p(a,t) = f(a,t), & (a,t) \in Q, \\ p(0,t) = \int_0^{a_\dagger} \beta(a,t)p(a,t)da, & t \in \mathbf{R}, \\ p(a,t) = p(a,t+T), & (a,t) \in Q, \end{cases} \tag{4.1}$$

where $Q = [0, a_\dagger) \times \mathbf{R}$.

This is the classical linear model for age-structured populations, where the vital rates β and μ are T-periodic with respect to the time t. Moreover, the population is subject to a T-periodic external flow f. The density of the population is supposed to be independent with respect to the spatial position (this is a model without diffusion). The biological relevance of (4.1) lies in that natural populations are usually subject to seasonal fluctuations.

Concerning the main assumptions on system (4.1) we suppose that the vital rates β and μ satisfy

(A1)
$$\begin{cases} \beta \in L^\infty(Q), \\ \beta(a,t) \geq 0 \ , \ \beta(a,t) = \beta(a,t+T) \quad \text{a.e. in } Q; \end{cases}$$

(A2)
$$\begin{cases} \mu(a,t) = \mu_0(a) + \overline{\mu}(a,t) \quad \text{a.e. in } Q, \\ \text{where } \mu_0 \in L^1_{loc}([0,a_\dagger)) \ , \ \mu_0(a) \geq 0 \quad \text{a.e. in } [0,a_\dagger), \\ \int_0^{a_\dagger} \mu_0(a)da = +\infty, \\ \overline{\mu} \in L^\infty(Q) \ , \ \overline{\mu}(a,t) \geq 0, \ \overline{\mu}(a,t) = \overline{\mu}(a,t+T) \quad \text{a.e. in } Q; \end{cases}$$

and the inflow rate f satisfies

(A3) $\qquad f \in L^\infty(Q) \ , \ f(a,t) = f(a,t+T) \quad \text{a.e. in } Q$.

This section is devoted mainly to the study of existence and uniqueness of solutions to system (4.1). We are interested in solutions belonging to the space

$$L^\infty_T(Q) = \{h \in L^\infty(Q); \ h(a,t) = h(a,t+T) \text{ a.e. in } Q\}$$

(endowed with the norm

$$\|h\|_\infty = \text{Ess Sup}_{(a,t)\in Q}|h(a,t)|) \ .$$

Namely, by *a solution to* (4.1), we mean a function $p \in L^\infty_T(Q)$, absolutely continuous along almost every characteristic line, which satisfies

$$\begin{cases} i) \quad Dp(a,t) = -\mu(a,t)p(a,t) + f(a,t), & \text{a.e. in } Q, \\ ii) \quad \lim_{\varepsilon\to 0+} p(\varepsilon, t+\varepsilon) = \int_0^{a_\dagger} \beta(a,t)p(a,t)da, & \text{a.e. } t \in \mathbf{R}, \\ iii) \quad p(a,t) = p(a,t+T) & \text{a.e. in } Q \ . \end{cases} \tag{4.2}$$

Since p is absolutely continuous along almost every characteristic line, (*ii*) is meaningful.

For biological reasons we are actually interested in nonnegative solutions to (4.1) so that we also consider the following requirement

$$p(a,t) \geq 0, \quad \text{a.e. in } Q \ . \tag{4.3}$$

However, for analysis of the optimal harvesting problem for periodic age-dependent population dynamics, which we will perform in the next chapter, we also need results with solutions that do not necessarily satisfy (4.3). Therefore we keep this condition separated from (4.2).

We now note that any p satisfying $(4.2)_i$ and $(4.2)_{ii}$, for $t > a_\dagger$ takes the form

$$p(a,t) = b(t-a)\Pi(a,t,a;\mu) + \int_0^a f(a-\sigma, t-\sigma)\Pi(a,t,\sigma;\mu)d\sigma \ , \tag{4.4}$$

where

$$\Pi(a,t,s;\mu) = \exp\Big\{ -\int_0^s \mu(a-\sigma, t-\sigma)d\sigma\Big\}, \tag{4.5}$$

$(a,t,s) \in [0,a_\dagger) \times \mathbf{R} \times [0,a_\dagger)$ and

$$b(t) = \lim_{\varepsilon\to 0+} p(\varepsilon, t+\varepsilon) \text{ a.e. } t \in \mathbf{R}$$

is a solution to the integral Volterra equation

$$b(t) = \int_0^t K(t,s;\mu)b(t-s)ds + F^\mu(t), \quad t \geq a_\dagger \ .$$

We have denoted by

$$K(t,s;\mu) = \begin{cases} \beta(s,t)\Pi(s,t,s;\mu), & \text{if } 0 \leq s \leq \min\{t, a_\dagger\}, \\ 0, & \text{elsewhere} \ , \end{cases}$$

$$
\begin{aligned}
F^{\mu}(t) &= \int_0^{\infty} \beta(a+t,t)p(a,0)\Pi(a+t,t,t;\mu)da \\
&+ \int_0^{\infty} \beta(a,t) \int_0^{\min\{t,a\}} \Pi(a,t,\sigma;\mu)f(a-\sigma,t-\sigma)d\sigma da \ ,
\end{aligned} \tag{4.6}
$$

where $t \geq 0$ and the functions β, Π, $p(a,0)$ are extended by zero outside their initial definition domains (see Section 2.1).

It is obvious that the first term of the right-hand side of (4.6) vanishes for $t > a_{\dagger}$, and thus any solution to (4.1) must necessarily have the form (4.4), with b being a T-periodic solution of the following equation

$$
b(t) = \int_0^{a_{\dagger}} K(t,s;\mu)b(t-s)ds + H^{\mu}(t), \quad t \in \mathbf{R} \ , \tag{4.7}
$$

where

$$
H^{\mu}(t) = \int_0^{a_{\dagger}} \beta(a,t) \int_0^{a} \Pi(a,t,\sigma;\mu)f(a-\sigma,t-\sigma)d\sigma \ da, \quad t \in \mathbf{R} \ . \tag{4.8}
$$

Since p must belong to $L_T^{\infty}(Q)$ we expect that b belongs to the space

$$
L_T^{\infty}(\mathbf{R}) = \{h \in L^{\infty}(\mathbf{R});\ h(t) = h(t+T) \text{ a.e. in } \mathbf{R}\} \ .
$$

LEMMA 2.4.1. *Let $b \in L_T^{\infty}(\mathbf{R})$ be a solution to (4.7). Then p, given by (4.4) (for any $t \in \mathbf{R}$) is a solution to (4.1). If (4.7) has at most one such solution, then (4.1) has at most one solution.*

Proof. It is obvious that if b is a solution to (4.7), then p given by (4.4) is a solution to (4.1). Moreover, since any solution to (4.1) must have the form (4.4) (for any $t \in \mathbf{R}$), with $b \in L_T^{\infty}(\mathbf{R})$ being a solution to (4.7), uniqueness follows from the uniqueness of the solution to this latter equation.

We are left with the problem of existence of a solution to (4.7). To this aim, for a fixed μ satisying **(A2)**, we define the linear and bounded operator

$$
\mathcal{A}^{\mu} : L_T^{\infty}(\mathbf{R}) \to L_T^{\infty}(\mathbf{R}) \ ,
$$

given by

$$
(\mathcal{A}^{\mu}h)(t) = \int_0^{a_{\dagger}} K(t,s;\mu)h(t-s)ds, \quad \forall h \in L_T^{\infty}(\mathbf{R}) \ . \tag{4.9}
$$

This definition is possible because $K(\cdot,\cdot;\mu) \in L^{\infty}(Q)$ and $K(t,s;\mu) = K(t+T,s;\mu)$ a.e. $(t,s) \in \mathbf{R} \times [0,a_{\dagger})$. As a consequence (4.7) can be written as the following abstract equation in $L_T^{\infty}(\mathbf{R})$

$$
b = \mathcal{A}^{\mu}b + H^{\mu} \ , \tag{4.10}
$$

with H^μ defined by (4.8).

If we denote by $r(\mathcal{A}^\mu)$ the spectral radius of the operator $\mathcal{A}^\mu$, we have as a consequence the following result:

THEOREM 2.4.2. *If* $r(\mathcal{A}^\mu) < 1$, *then equation (4.7) has a unique solution in* $L_T^\infty(\mathbf{R})$.

Remark that if we denote by

$$\overline{\beta}(a) = \text{Ess sup}_{t\in\mathbf{R}}\beta(a,t) \quad \text{a.e. } a \in [0, a_\dagger)\ ,$$

then (4.9) allows us to conclude that

$$r(\mathcal{A}^\mu) \le \int_0^{a_\dagger} \overline{\beta}(s) e^{-\int_0^a \mu_0(\sigma)d\sigma} da\ .$$

So, if $\int_0^{a_\dagger} \overline{\beta}(s) e^{-\int_0^a \mu_0(\sigma)d\sigma} da < 1$ (which means that the fertility rate is "dominated" by the mortality rate; see also the previous sections for the biological significance of this condition), then the spectral radius satisfies

$$r(\mathcal{A}^\mu) < 1\ .$$

In the case of time-independent vital rates, the operator $\mathcal{A}^\mu$ becomes the convolution operator

$$(\mathcal{A}^\mu h)(t) = \int_0^{a_\dagger} \beta(s) e^{-\int_0^s \mu(\sigma)d\sigma} h(t-s)ds, \quad t \in \mathbf{R}$$

and we have

$$\|(\mathcal{A}^\mu)^k\|_{L(L_T^\infty(\mathbf{R}))} = \|(\mathcal{A}^\mu)\|^k_{L(L_T^\infty(\mathbf{R}))}, \quad \forall k \in \mathbf{N}^*\ .$$

Thus condition $r(\mathcal{A}^\mu) < 1$ becomes

$$r(\mathcal{A}^\mu) = \|\mathcal{A}^\mu\|_{L(L_T^\infty(\mathbf{R}))} = \int_0^{a_\dagger} \beta(s) e^{-\int_0^s \mu(\sigma)d\sigma} ds < 1\ .$$

From a biological point of view, this condition means that the net rate of the population (see Section 2.3) being less than 1, the population would go extinct if the external supply f is null.

Since we have to consider problem (4.1) with different choices for μ (satisfying **(A2)**), we need to state the following comparison result:

LEMMA 2.4.3. *Let μ_1, μ_2 satisfy* **(A2)** *and $\mu_1(a,t) \leq \mu_2(a,t)$ a.e. in Q. Then*

$$r(\mathcal{A}^{\mu_1}) \geq r(\mathcal{A}^{\mu_2}) . \tag{4.11}$$

Proof. Since $K(t,s;\mu) \geq 0$, then for any μ satisfying **(A2)** we have

$$\|(\mathcal{A}^{\mu})^k\|_{L(L_T^\infty(\mathbf{R}))} = \int_{[0,a_\dagger]^k} K(t,s_1;\mu)K(t-s_1,s_2;\mu)$$

$$\cdots K(t-s_1-s_2-\cdots s_{k-1},s_k;\mu)ds_1\cdots ds_k ,$$

for $k = 1,2,\ldots$. Thus, from the definition of $K(t,s,\mu)$ and (4.5) we get

$$\|(\mathcal{A}^{\mu_2})^k\|_{L(L_T^\infty(\mathbf{R}))} \leq \|(\mathcal{A}^{\mu_1})^k\|_{L(L_T^\infty(\mathbf{R}))}$$

and (4.11) follows.

We finally have the following result:

THEOREM 2.4.4. *Suppose that $r(\mathcal{A}^{\mu}) < 1$. Then problem (4.1) has a unique solution $p^{\mu} \in L_T^\infty(Q)$. Moreover we have*

(1) *if $f(a,t) \geq 0$ a.e. in Q, then $p^{\mu}(a,t) \geq 0$ a.e. in Q;*

(2) *if $f(a,t) > 0$ a.e. in Q, then $p^{\mu}(a,t) > 0$ a.e. in Q;*

(3) *if μ_1, μ_2 satisfy* **(A2)** *and $\mu_1(a,t) \leq \mu_2(a,t)$ a.e. in Q and if $f(a,t) \geq 0$ a.e. in Q, then*

$$p^{\mu_2}(a,t) \leq p^{\mu_1}(a,t) \quad \text{a.e. in } Q;$$

(4) *if $f_n \to f$ in $L_T^\infty(Q)$, then*

$$p_n^{\mu} \to p^{\mu}$$

in $L_T^\infty(Q)$, where p_n^{μ}, p^{μ} are the solutions of (4.1) corresponding to $f := f_n$ and f, respectively.

Proof. For any μ satisfying **(A2)**, the existence and uniqueness of the solution p^{μ} of (4.1) follow by Theorem 2.4.2 and Lemma 2.4.1. We note that if $f(a,t) \geq 0$ a.e. in Q, then $H^{\mu}(t) \geq 0$ a.e. in $\mathbf{R}$. The solution b of (4.10) is the limit in $L_T^\infty(\mathbf{R})$ of the iterative sequence

$$\begin{cases} b_0(t) = H^{\mu}(t), & t \in \mathbf{R}, \\ b_{n+1}(t) = H^{\mu}(t) + \displaystyle\int_0^{a_\dagger} K(t,s;\mu)b_n(t-s)ds, & t \in \mathbf{R}, \end{cases} \tag{4.12}$$

$n \in \mathbf{N}^*$, and since $K(t,s;\mu) \geq 0$, a.e. $(t,s) \in \mathbf{R} \times [0,a_\dagger)$ and $H^\mu(t) \geq 0$, a.e. $t \in \mathbf{R}$, we have $b_n(t) \geq 0$, a.e. $t \in \mathbf{R}$ and consequently $b(t) \geq 0$, a.e. $t \in \mathbf{R}$. Thus by (4.4) we have $p^\mu(a,t) \geq 0$ a.e. in Q.

If $f(a,t) > 0$ a.e. in Q, then $H^\mu(t) > 0$ a.e. $t \in \mathbf{R}$ and in the same manner as above it follows that $p^\mu(a,t) > 0$ a.e. in Q.

Concerning (3), if we denote by b_n^1 and b_n^2 the approximating sequences in (4.12) relative to μ_1 and μ_2, respectively, and since

$$K(t,s;\mu_2) \leq K(t,s;\mu_1) \quad \text{a.e. } (t,s) \in \mathbf{R} \times [0,a_\dagger) ,$$

$$H^{\mu_2}(t) \leq H^{\mu_1}(t) \quad \text{a.e. } t \in \mathbf{R} ,$$

we have $b_n^2(t) \leq b_n^1(t)$, a.e. $t \in \mathbf{R}$. Passing to the limit $(n \to +\infty)$, condition (3) follows via (4.4).

Finally, if $f_n \to f$ in $L_T^\infty(Q)$, then $H_n^\mu \to H^\mu$ in $L_T^\infty(\mathbf{R})$ (where H_n^μ and H^μ are given by (4.8) for $f := f_n$ and f, respectively). Thus by (4.7) we have

$$b_n^\mu \to b^\mu \quad \text{in } L_T^\infty(\mathbf{R}),$$

where b_n^μ and b^μ are the solutions of (4.7) corresponding to $f := f_n$ and f, respectively and (4) follows by (4.4).

Before ending this section, we note that, by (4.2) the solution p to problem (4.1) also satisfies

$$\begin{aligned} &\int_0^T \int_0^{a_\dagger} \big[-D\varphi + \mu(a,t)\varphi - \beta(a,t)\varphi(0,t) \big] p(a,t)da\ dt \\ = &\int_0^T \int_0^{a_\dagger} f(a,t)\varphi(a,t)da\ dt , \end{aligned} \tag{4.13}$$

where φ is any absolutely continuous function along almost every characteristic line, satisfying

$$\begin{cases} \varphi \in L_T^\infty(Q), \\ D\varphi \in L^1((0,a_\dagger) \times (0,T)), \\ \mu\varphi \in L^1((0,a_\dagger) \times (0,T)) . \end{cases} \tag{4.14}$$

By (4.14) we conclude that

$$\varphi(a_\dagger, t) = \lim_{\varepsilon \to 0^+} \varphi(a_\dagger - \varepsilon, t - \varepsilon) \quad \text{a.e. in } \mathbf{R}$$

and thus (4.13) follows from (4.2), integrating by parts.

Denote by Φ the set of all test functions φ with the above mentioned properties.

We call a *weak solution* to problem (4.1) any function $p \in L^\infty_T(Q)$ satisfying (4.13) for any $\varphi \in \Phi$.

Actually we have the following uniqueness result:

THEOREM 2.4.5. *Suppose that $r(\mathcal{A}^\mu) < 1$. Then system (4.1) has at most one weak solution.*

Proof. If $p^1 \in L^\infty_T(Q)$ and $p^2 \in L^\infty_T(Q)$ are two weak solutions to (4.1), then $w = p^1 - p^2$ satisfies

$$\int_0^T \int_0^{a_\dagger} \big[- D\varphi + \mu(a,t)\varphi - \beta(a,t)\varphi(0,t) \big] w(a,t) da\, dt = 0 \,, \tag{4.15}$$

for any $\varphi \in \Phi$.

Let us consider the problem

$$\begin{cases} (D\varphi - \mu\varphi)(a,t) + \beta(a,t)\varphi(0,t) = sign\ w(a,t) & \text{a.e. in } Q, \\ \varphi(a,t) = \varphi(a,t+T) & \text{a.e. in } Q, \\ \varphi(a_\dagger,t) = 0 & \text{a.e. in } \mathbf{R}\,, \end{cases} \tag{4.16}$$

where we seek a solution $\varphi \in \Phi$ satisfying

$$\varphi(0,t) = \lim_{\varepsilon \to 0+} \varphi(\varepsilon, t+\varepsilon), \quad \varphi(a_\dagger,t) = \lim_{\varepsilon \to 0+} \varphi(a_\dagger - \varepsilon, t-\varepsilon) \quad \text{a.e. in } \mathbf{R}\,.$$

Indeed, (4.16) has the solution

$$\begin{aligned} \varphi(a,t) = \int_a^{a_\dagger} & e^{-\int_a^\sigma \mu(\rho,t-a+\rho)d\rho} \\ & \cdot \big[\beta(\sigma,t-a+\sigma)X(t-a+\sigma) - sign\ w(\sigma,t-a+\sigma)\big] d\sigma \,, \end{aligned} \tag{4.17}$$

$a \in [0,a_\dagger)$, $t \in \mathbf{R}$, where $X \in L^\infty_T(\mathbf{R})$ is the solution of the equation

$$X(t) = \int_0^{a_\dagger} K(t+\sigma,\sigma;\mu)X(t+\sigma)d\sigma + Y(t), \quad t \in \mathbf{R}\,, \tag{4.18}$$

with

$$Y(t) = -\int_0^{a_\dagger} \Pi(\sigma,t+\sigma,\sigma;\mu) sign\ w(\sigma,t+\sigma)d\sigma, \quad t \in \mathbf{R}\,.$$

We claim that (4.18) has a unique solution $X \in L_T^\infty(\mathbf{R})$. In fact, since $Y \in L_T^\infty(\mathbf{R})$ we can write (4.18) as the abstract equation

$$X = \mathcal{A}_0^\mu X + Y \ ,$$

where $\mathcal{A}_0^\mu : L_T^\infty(\mathbf{R}) \to L_T^\infty(\mathbf{R})$ is the bounded operator defined as

$$(\mathcal{A}_0^\mu h)(t) = \int_0^{a_\dagger} K(t+\sigma, \sigma; \mu) h(t+\sigma) d\sigma \ .$$

Now $\mathcal{A}_0^\mu$ is the restriction to $L_T^\infty(Q)$ of the adjoint of the operator $\mathcal{A}^\mu$ defined in (4.9). Thus $r(\mathcal{A}_0^\mu) < 1$ and (4.18) has a unique solution.

Now the function φ defined in (4.17) satisfies (4.14). In fact the first two conditions of (4.14) can be directly checked, while for the third condition it is enough to note that

$$|\mu_0(a)\varphi(a,t)| \le C \int_a^{a_\dagger} \mu_0(a) e^{-\int_a^\sigma \mu_0(\rho)d\rho} d\sigma \quad \text{a.e. in } Q \ ,$$

where $C \in \mathbf{R}_+^*$ and consequently

$$\int_0^{a_\dagger} \mu_0(a)|\varphi(a,t)| da \le C \int_0^{a_\dagger} d\sigma \int_0^\sigma \mu_0(a) e^{-\int_a^\sigma \mu_0(\rho)d\rho} da < +\infty \text{ a.e. } t \in \mathbf{R} \ .$$

We can substitute φ in (4.15) and obtain

$$\int_0^T \int_0^{a_\dagger} |w(a,t)| da \ dt = 0 \ .$$

Thus $w(a,t) = 0$ a.e. in Q and $p_1^\mu(a,t) = p_2^\mu(a,t)$ a.e. in Q.

We note that in the proof of the previous theorem we solved a special case of the following problem:

$$\begin{cases} Dq - \mu(a,t)q + \beta(a,t)q(0,t) = f(a,t), & (a,t) \in Q, \\ q(a_\dagger, t) = 0, & t \in \mathbf{R}, \\ q(a,t) = q(a, t+T), & (a,t) \in Q \ . \end{cases} \tag{4.19}$$

This is a dual problem with respect to (4.1). Using the same argument used for solving (4.16) we can state the following theorem:

THEOREM 2.4.6. *Suppose that* $r(\mathcal{A}^\mu) < 1$. *Then problem (4.19) has one and only one solution* q.

If moreover $f(a,t) \ge 0$ *a.e. in* Q, *then* $q(a,t) \le 0$ *a.e. in* Q.

Again, by a *solution* to (4.19) we mean a function q, absolutely continuous along almost every characteristic line, satisfying (4.14) and such that

$$\begin{cases} Dq(a,t) = \mu(a,t)q(a,t) - \beta(a,t)q(0,t) + f(a,t) & \text{a.e. in } Q, \\ q(a_\dagger,t) = \lim_{h\to 0+} q(a_\dagger - h, t-h) = 0 & \text{a.e. in } \mathbf{R}, \end{cases}$$

where $q(0,t) = \lim_{h\to 0+} q(h,t+h)$ a.e. in $\mathbf{R}$.

*

* *

The linear theory of continuous age-dependent population dynamics started in 1911 with the work of A. Lotka and F.R. Sharpe [80] (see also [67]) and with the paper of A.G. McKendrick [70], who emphasized the connection with the renewal equation. The study of the basic properties of the solution to the linear model is related to the theory of Volterra integral equations (see Feller [37]). The first attention to the problem with time dependent vital rates was given by H. Langhaar [61].

Some of the first nonlinear continuous age-dependent models were proposed by M.E. Gurtin and R.C. MacCamy [41], [43] and by F. Hoppensteadt [50]. For the study of some separable models we refer to S. Busenberg and M. Iannelli [27] and S. Aniţa [10]. An abstract formulation of the nonlinear model was developed and treated by G. Webb [85], [86], [87] and by M. Kubo [56]. An extended treatment of both the modeling aspects and the mathematical methods has been provided by J. Metz and O. Diekmann [72].

The study of time periodic models is of great importance due to the fact that often the vital rates and the inflow are time periodic. In this direction we mention the very recent papers of M. Kubo and M. Langlais [57] and of S. Aniţa, M. Iannelli, M.-Y. Kim and E.-J. Park [11]. For related results concerning the model with time-periodic vital rates we refer to Thieme [82].

The approach in Sections 2.1 and 2.2 is classical. We also refer to the monographs of J. Impagliazzo [53] and of M. Iannelli [52].

The study of the asymptotic behaviour of the solution to the linear model is based on classical Laplace transforms techniques (see [34] and [52]). The results concerning the nonlinear model in Section 2.3 is based on the asymptotic behaviour in the linear case.

The results in Section 2.4 were established by S. Aniţa, M. Iannelli, M.-Y. Kim and E.-J. Park [11]. For a model involving the space variable and time-periodic vital rates and inflow we refer to [57].

2.5. Exercises

1. Solve the problem

$$Dp(a,t) + p(a,t) = 1, \qquad (a,t) \in (0,1) \times (0,+\infty),$$

$$p(0,t) = \int_0^1 p(a,t)da, \qquad t \in (0,+\infty),$$

$$p(a,0) = 1, \qquad a \in (0,1).$$

Hint. We follow the steps in the proof of Theorem 2.1.1. We look for p of the form

$$p(a,t) = \begin{cases} e^{-t} + \displaystyle\int_0^t e^{-t+s} ds, & \text{if } t < a, \\ e^{-a} b(t-a) + \displaystyle\int_0^a e^{-a+s} ds, & \text{if } t > a, \end{cases}$$

where $b(t) = \int_0^1 p(a,t)da$ a.e. $t \in (0,+\infty)$; b is the solution of

$$b(t) = \begin{cases} \displaystyle\int_0^t \left(e^{-a} b(t-a) + 1 - e^{-a}\right) da + \int_t^1 da & \text{if } t < 1, \\ \displaystyle\int_0^1 \left(e^{-a} b(t-a) + 1 - e^{-a}\right) da & \text{if } t > 1. \end{cases}$$

Solving this last equation (which is a Volterra equation) we get $b(t) \equiv 1$ and consequently we obtain $p(a,t) \equiv 1$.

2. Solve the problem

$$Dp(a,t) + \frac{1}{1-a} p(a,t) = 1 - a, \qquad (a,t) \in (0,1) \times (0,+\infty),$$

$$p(0,t) = 2\int_0^1 p(a,t)da, \qquad t \in (0,+\infty),$$

$$p(a,0) = 0, \qquad a \in (0,1).$$

Hint. Following the same steps as in the previous exercise we get the solution $p(a,t) = (1-a)t$, $(a,t) \in [0,1] \times [0,+\infty)$.

3. Find the solution of

$$Dp(a,t) + \frac{1}{1-a}p(a,t) = 0, \qquad (a,t) \in (0,1) \times (0,+\infty),$$

$$p(0,t) = 2\int_0^1 p(a,t)da, \qquad t \in (0,+\infty),$$

$$p(a,0) = 1-a, \qquad a \in (0,1).$$

Hint. The solution is $p(a,t) = 1-a$, $(a,t) \in [0,1] \times [0,+\infty)$.

4. Determine the solution of the following age-dependent population dynamics

$$Dp(a,t) + \left[\frac{1}{(a_\dagger - a)^2} + t\right] p(a,t) = 0, \qquad (a,t) \in (0,a_\dagger) \times (0,+\infty),$$

$$p(0,t) = \int_0^{a_\dagger} \frac{1}{(a_\dagger - a)^2} p(a,t)da, \qquad t \in (0,+\infty),$$

$$p(a,0) = e^{-\frac{1}{a_\dagger - a}}, \qquad a \in (0,a_\dagger).$$

Hint. Following the steps in the proof of Theorem 2.1.1 we obtain that

$$p(a,t) = e^{-\frac{1}{a_\dagger - a} - \frac{t^2}{2}}, \qquad (a,t) \in [0,a_\dagger] \times [0,+\infty).$$

5. Show that if assumptions **(A1)-(A3)** in Section 2.1 hold and if in addition

$$\beta(a,t) > 0 \qquad \text{a.e. } (a,t) \in (a_1,a_2) \times (0,+\infty),$$

$$p_0(a) > 0 \qquad \text{a.e. } a \in (a_1,a_2)$$

(where $0 \le a_1 < a_2 \le a_\dagger$), then the solution to (1.1) in Section 2.1 satisfies

$$p(a,t) > 0 \qquad \text{a.e. } a \in (0,a_\dagger),$$

for t large enough.

Hint. The assumptions on β and p_0 imply that

$$p(0,t) = \int_0^{a_\dagger} \beta(a,t)p(a,t)da > 0$$

for $t \in (0, a_2 - a_1)$ and consequently $p(a,t) > 0$ for $t - a \in (0, a_2 - a_1)$. Using again the renewal law we get that $p(0,t) > 0$ for $t \in (a_1, 2a_2 - a_1)$. Repeating this procedure we obtain that $p(0,t) > 0$ for t large enough,and so we get the conclusion.

6. Show that under assumptions **(A1)-(A3)** in Section 2.1, the solution of

$$Dp(a,t) + \mu(a,t)p(a,t) = -\frac{1}{T-t}p(a,t), \qquad (a,t) \in (0,a_\dagger) \times (0,T),$$

$$p(0,t) = \int_0^{a_\dagger} \beta(a,t)p(a,t)da, \qquad t \in (0,T),$$

$$p(a,0) = p_0(a), \qquad a \in (0,a_\dagger)$$

satisfies

$$p(a,T) = 0 \qquad \text{a.e. } a \in (0,a_\dagger).$$

Here $p(a,T)$ should be seen as the limit $\lim_{\varepsilon \to 0^+} p(a-\varepsilon, T-\varepsilon)$.

REMARK. This is a null controllability result for linear age structured population dynamics and the term $-\frac{1}{T-t}p(a,t)$ represents a feedback control.

Hint. The solution p is given by

$$p(a,t) = e^{-\int_0^t \frac{1}{T-s}ds}\tilde{p}(a,t), \qquad (a,t) \in (0,a_\dagger) \times (0,T),$$

where $\tilde{p}$ is the solution to

$$Dp(a,t) + \mu(a,t)p(a,t) = 0, \qquad (a,t) \in (0,a_\dagger) \times (0,T),$$

$$p(0,t) = \int_0^{a_\dagger} \beta(a,t)p(a,t)da, \qquad t \in (0,T),$$

$$p(a,0) = p_0(a), \qquad a \in (0,a_\dagger).$$

Taking $t \to T$ we get the conclusion.

7. Solve the problem

$$Dp + \frac{1}{1-a}p + \left(\int_0^1 p(a,t)da\right)p(a,t) = 0, \qquad (a,t) \in (0,1) \times (0,+\infty),$$

$$p(0,t) = 2\int_0^1 p(a,t)da, \qquad t \in (0,+\infty),$$

$$p(a,0) = 1-a, \qquad a \in (0,1).$$

Hint. This is a "separable" problem (see Section 2.2) and has a unique solution p, given by

$$p(a,t) = x(t)\tilde{p}(a,t), \quad (a,t) \in [0,1] \times \mathbf{R}^+,$$

where $\tilde{p}$ is the solution of Exercise 3 and x is the solution of

$$x'(t) + \left(\int_0^1 \tilde{p}(a,t)da\right) x^2(t) = 0, \qquad t \in (0,+\infty),$$

$$x(0) = 1.$$

Since $\tilde{p}(a,t) = 1 - a$, $(a,t) \in [0,1] \times \mathbf{R}^+$, we may conclude that x is the solution of

$$x'(t) + \frac{1}{2}x^2(t) = 0, \qquad t \in (0,+\infty),$$

$$x(0) = 1.$$

By a simple calculation we get $x(t) = \dfrac{2}{2+t}$, $t \in \mathbf{R}^+$ and in conclusion

$$p(a,t) = \frac{2(1-a)}{2+t}, \quad (a,t) \in [0,1] \times \mathbf{R}^+$$

is the unique solution of the problem.

8. Investigate the large time behaviour of the solution of the following problem:

$$Dp + p + \left(\int_0^1 p(a,t)da\right) p(a,t) = 0, \qquad (a,t) \in (0,1) \times (0,+\infty),$$

$$p(0,t) = \int_0^1 p(a,t)da, \qquad t \in (0,+\infty),$$

$$p(a,0) = 1, \qquad a \in (0,1).$$

Hint. Arguing as in Section 2.3 we conclude that

$$\lim_{t\to+\infty} \|p(t)\|_{L^\infty(0,1)} = 0.$$

9. If β, μ and f satisfy assumptions **(A1)**-**(A2)** in Section 2.3 and **(A3)** in Section 2.4 and if

$$\int_0^{a_\dagger} \beta(a) e^{-\int_0^a \mu(s)ds} da > 1,$$

$$f(a,t) > 0 \qquad \text{a.e. } (a,t) \in (0, a_\dagger) \times \mathbf{R},$$

then the periodic problem (4.1) has no nonnegative solution.

Hint. If there exists a nonnegative solution p of (4.1), then $p(a,t) > 0$ (is strictly positive because f is strictly positive). By Remark 2.3.2 we conclude that

$$\|p(t)\|_{L^1(0,a_\dagger)} \to +\infty$$

as $t \to +\infty$. Since p is T-periodic, the last convergence is not possible.

10. Solve the problem

$$\begin{aligned} &Dp(a,t) + p(a,t) = 1 + \cos t, && (a,t) \in (0,1) \times \mathbf{R}, \\ &p(0,t) = \int_0^1 p(a,t)da, && t \in \mathbf{R}, \\ &p(a,t) = p(a, t + 2\pi), && (a,t) \in (0,1) \times \mathbf{R}. \end{aligned}$$

Hint. We follow the steps in Section 2.4. We look for p of the form

$$p(a,t) = e^{-a} b(t-a) + \int_0^a e^{-(a-s)} (1 + \cos(t - a + s))\, ds,$$

$(a,t) \in [0,1] \times \mathbf{R}$, where

$$b(t) = \int_0^1 p(a,t)da \qquad \text{a.e. } t \in \mathbf{R}.$$

CHAPTER 3

OPTIMAL CONTROL OF POPULATION DYNAMICS

This chapter concerns some of the most important optimal control problems related to age-dependent population dynamics. The main goal is to prove the existence of an optimal control and to obtain first order necessary conditions of optimality. These conditions allow calculation or approximation of the optimal control and the optimal value of the cost functional. We shall also present a fractional step scheme for a certain nonlinear problem. This scheme gives a very good method of approximating the optimal control.

Throughout this chapter $a_\dagger$, $T \in (0, +\infty)$, $Q_T = (0, a_\dagger) \times (0, T)$ and Dp is the directional derivative

$$Dp(a,t) = \lim_{\varepsilon \to 0} \frac{p(a+\varepsilon, t+\varepsilon) - p(a,t)}{\varepsilon}.$$

A quite general optimal control problem can be formulated in the following form:

$$Minimize\ \mathcal{L}(u, x^u),$$

subject to $u \in K$, where x^u is the solution to

$$F(u,x) = 0. \tag{1}$$

Here U, X, Y are real Banach spaces (with duals U^*, X^* and Y^*, respectively), $K \subset U$ is a closed subset and $\mathcal{L} : U \times X \to \mathbf{R}$, $F : U \times X \to Y$ are given functions. U is the set of the controllers.

Assume that for any $u \in K$, the equation (1) admits a unique solution, denoted by x^u.

We say that the control $u^* \in K$ is *optimal* if

$$\mathcal{L}(u^*, x^{u^*}) \le \mathcal{L}(u, x^u),$$

for any $u \in K$. The pair (u^*, x^{u^*}) is an *optimal pair* and $\mathcal{L}(u^*, x^{u^*})$ is the *optimal value* of the cost functional.

We assume that $\mathcal{L}$ and F are smooth enough and that the optimal control problem has at least one optimal control.

The function $u \mapsto x^u$ is Gâteaux differentiable and consider $z = dx^{u^*}(v)$, where $v \in U$ and satisfies

$$u^* + \varepsilon v \in K \quad \text{for any } \varepsilon > 0 \text{ small enough.} \tag{2}$$

Denote by $\mathcal{V}$ the set of all $v \in U$ which satisfies (2).

z is the solution to

$$d_u F(u^*, x^{u^*})v + d_x F(u^*, x^{u^*})z = 0. \tag{3}$$

Here, $d_u F(\cdot, x)$ is the Gâteaux differential of $u \mapsto F(u, x)$ $(x \in X)$ and $d_x F(u, \cdot)$ is the Gâteaux differential of $x \mapsto F(u, x)$ $(u \in U)$.

Since

$$\mathcal{L}(u^*, x^{u^*}) \leq \mathcal{L}(u^* + \varepsilon v, x^{u^* + \varepsilon v}), \quad \forall v \in \mathcal{V},$$

for any $\varepsilon > 0$ small enough, we conclude that

$$\left(v, d_u \mathcal{L}(u^*, x^{u^*})\right) + \langle z, d_x \mathcal{L}(u^*, x^{u^*}) \rangle \geq 0. \tag{4}$$

We have denoted by $(\cdot, \cdot)$, $\langle \cdot, \cdot \rangle$ the duality between U, U^* and X, X^*, respectively.

Assume that the dual equation

$$\left(d_x F(u^*, x^{u^*})\right)^* p = d_x \mathcal{L}(u^*, x^{u^*}) \tag{1$'$}$$

has a solution p. By (3) and (1)$'$ we obtain that

$$-\left(v, \left(d_u F(u^*, x^{u^*})\right)^* p\right) = \langle z, d_x \mathcal{L}(u^*, x^{u^*}) \rangle = 0. \tag{5}$$

By (4) and (5) we see that

$$\left(v, \left(d_u F(u^*, x^{u^*})\right)^* p - d_u \mathcal{L}(u^*, x^{u^*})\right) \leq 0$$

for any $v \in \mathcal{V}$. This implies that

$$\left(d_u F(u^*, x^{u^*})\right)^* p - d_u \mathcal{L}(u^*, x^{u^*}) \in N_K(u^*). \tag{6}$$

Equations (1), (1)$'$ and (6) represent the maximum principle (Pontriagin's principle) for the optimal control problem.

If F and $\mathcal{L}$ are not smooth, then we must "approximate" the initial problem by the following sequence of optimal control problems:

$$Minimize\ \mathcal{L}_\varepsilon(u, x_\varepsilon^u) + \|u - u^*\|_U^2$$

($\varepsilon > 0$), subject to $u \in K$, where x_ε^u is the solution to

$$F_\varepsilon(u, x) = 0.$$

Here $\mathcal{L}_\varepsilon$ and F_ε are smooth aproximations of $\mathcal{L}$ and F, respectively (see e.g., Appendix 1 for approximations for convex functions and for locally Lipschitz functions).

The idea is now to pass to the limit (using a priori estimates) into the optimality system (1)′ and (6), corresponding to $\mathcal{L} := \mathcal{L}_\varepsilon$ and $F := F_\varepsilon$.

This is only a general scheme and not a rigourous proof for the maximum principle.

3.1. Optimal harvesting for linear age-dependent population dynamics

This section is devoted to the optimal harvesting problem governed by linear age-dependent population dynamics. The main goal of this section is to obtain first order necessary conditions of optimality. These conditions allow us to calculate the optimal control and the optimal harvest.

Namely we deal with the following problem:

$$\textbf{(OH)} \qquad \textit{Maximize} \int_0^T \int_0^{a_\dagger} u(a,t)p^u(a,t)da\, dt,$$

subject to $u \in \mathcal{U} = \{h \in L^\infty(Q_T);\ \zeta_1(a,t) \le h(a,t) \le \zeta_2(a,t) \text{ a.e. in } Q_T\}$, where p^u is the solution of

$$\begin{cases} Dp(a,t) + \mu(a,t)p(a,t) = f(a,t) - u(a,t)p(a,t) & (a,t) \in Q_T, \\ p(0,t) = \displaystyle\int_0^{a_\dagger} \beta(a,t)p(a,t)da & t \in (0,T), \\ p(a,0) = p_0(a) & a \in (0, a_\dagger). \end{cases} \tag{1.1}$$

Here $\zeta_1, \zeta_2 \in L^\infty(Q_T)$ $0 \le \zeta_1(a,t) \le \zeta_2(a,t)$ a.e. in Q_T. $\mathcal{U}$ is the set of controls; $u \in \mathcal{U}$ is the control (or the harvesting effort). The integral $\displaystyle\int_0^T \int_0^{a_\dagger} u(a,t)p^u(a,t)da\, dt$ gives the total harvest.

Problem (**OH**) asks us to find the control which gives the best (maximal) harvest. Such a control is called "optimal control" (or "optimal effort").

Assume that the following hypotheses hold throughout this section:

(**A1**) $\qquad \beta \in L^\infty(Q_T)$, $\beta(a,t) \ge 0$ a.e. in Q_T;

$$
\textbf{(A2)} \qquad \begin{cases} \mu \in L^1_{loc}([0,a_\dagger) \times [0,T]) \;,\; \mu(a,t) \geq 0 \quad \text{a.e. in } Q_T, \\ \displaystyle\int_0^{a_\dagger} \mu(a, t - a_\dagger + a) da = +\infty, \\ \text{where } \mu(a,t) \text{ is extended by zero on } (0,a_\dagger) \times (-\infty, 0); \end{cases}
$$

$$
\textbf{(A3)} \qquad \begin{cases} p_0 \in L^2(0,a_\dagger) \;,\; p_0(a) \geq 0 \quad & \text{a.e. in } (0,a_\dagger), \\ f \in L^2(Q_T) \;,\; f(a,t) \geq 0 \quad & \text{a.e. in } Q_T. \end{cases}
$$

Since $\mu := \mu + u$ satisfies **(A2)** (for any $u \in \mathcal{U}$), we conclude via Theorem 2.1.1 that system (1.1) has a unique solution $p^u \in L^\infty(0,T;L^1(0,a_\dagger))$; this solution is nonnegative. Since in addition $p_0 \in L^2(0,a_\dagger)$ and $f \in L^2(Q_T)$, then Remark 2.1.2 allows us to conclude that $p^u \in L^\infty(0,T;L^2(0,a_\dagger))$, for any $u \in \mathcal{U}$.

EXISTENCE OF OPTIMAL CONTROL

THEOREM 3.1.1. *Problem* **(OH)** *admits at least one optimal control.*

Proof. Define

$$
\Psi(u) = \int_0^T \int_0^{a_\dagger} u(a,t) p^u(a,t) da\, dt, \quad u \in \mathcal{U}
$$

and let

$$
d = \sup_{u \in \mathcal{U}} \Psi(u).
$$

Using the comparison result in Section 2.1 (Theorem 2.1.2) we obtain

$$
0 \leq \Psi(u) \leq \|\zeta_2\|_{L^\infty(Q_T)} \int_0^T \int_0^{a_\dagger} p^0(a,t) da\, dt < +\infty
$$

(p^0 is the solution of (1.1) corresponding to $u := 0$), for any $u \in \mathcal{U}$ and consequently $d \in [0, +\infty)$.

Let $\{u_n\}_{n \in \mathbf{N}^*} \subset \mathcal{U}$ be such that

$$
d - \frac{1}{n} < \Psi(u_n) \leq d. \tag{1.2}
$$

The same comparison result allows us to conclude that

$$
0 \leq p^{u_n}(a,t) \leq p^0(a,t) \quad \text{a.e. in } Q_T
$$

and so, on a subsequence, also denoted by $\{p^{u_n}\}$, we have

$$p^{u_n} \to p^* \quad \text{weakly in } L^2(Q_T).$$

We recall the Mazur theorem (see [21]) and a corollary, which will be used often throughout this chapter

THEOREM (Mazur). *A convex subset in a real Banach space is weakly closed if and only if it is strongly convex.*

As a consequence we obtain the following

COROLLARY (Mazur). *Let $\{x_n\}_{n\in\mathbf{N}}$ be a sequence in a real Banach space X, which is weakly convergent to $x \in X$. Then there exists a sequence $\{y_n\}_{n\in\mathbf{N}} \subset X$, $y_n \in conv\ \{x_k;\ k \geq n+1\}$, $\forall n \in \mathbf{N}$, such that $\{y_n\}_{n\in\mathbf{N}}$ converges strongly to x.*

Using now Mazur's theorem we obtain a sequence $\{\tilde{p}_n\}$ satisfying

$$\tilde{p}_n = \sum_{i=n+1}^{k_n} \lambda_i^n p^{u_i}\ ,\ \lambda_i^n \geq 0\ ,\ \sum_{i=n+1}^{k_n} \lambda_i^n = 1$$

$(k_n \geq n+1)$ and

$$\tilde{p}_n \to p^* \quad \text{in } L^2(Q_T). \tag{1.3}$$

Let the controls $\tilde{u}_n$ be defined as follows

$$\tilde{u}_n(a,t) = \begin{cases} \dfrac{\sum_{i=n+1}^{k_n} \lambda_i^n p^{u_i}(a,t)u_i(a,t)}{\sum_{i=n+1}^{k_n} \lambda_i^n p^{u_i}(a,t)}, & \text{if } \sum_{i=n+1}^{k_n} \lambda_i^n p^{u_i}(a,t) \neq 0, \\ \zeta_1(a,t), & \text{if } \sum_{i=n+1}^{k_n} \lambda_i^n p^{u_i}(a,t) = 0. \end{cases}$$

These controls satisfy $\tilde{u}_n \in \mathcal{U}$ and

$$\tilde{p}_n(a,t) = p^{\tilde{u}_n}(a,t) \quad \text{a.e. in } Q_T.$$

We can take a subsequence (also denoted by $\{\tilde{u}_n\}$) such that

$$\begin{cases} \tilde{u}_n \to u^* & \text{weakly in } L^2(Q_T), \\ \Psi(\tilde{u}_n) = \displaystyle\sum_{i=n+1}^{k_n} \lambda_i^n \Psi(u_i) \to d & \text{as } n \to +\infty \end{cases} \tag{1.4}$$

(we have used (1.2)). Thus by (1.3) and (1.4) we infer that

$$d = \lim_{n \to +\infty} \Psi(\tilde{u}_n) = \int_0^T \int_0^{a_\dagger} u^*(a,t) p^*(a,t) da\, dt. \tag{1.5}$$

We are left with the proof of the following equality:

$$p^*(a,t) = p^{u^*}(a,t) \quad \text{a.e. in } Q_T. \tag{1.6}$$

To prove (1.6) we recall that $p^{\tilde{u}_n}$ is a weak solution of (1.1) corresponding to $u := \tilde{u}_n$ (see Section 2.1), that is:

$$\int_0^T \int_0^{a_\dagger} [-D\varphi(a,t) + (\mu(a,t) + \tilde{u}_n(a,t))\varphi(a,t) - \beta(a,t)\varphi(0,t)] p^{\tilde{u}_n}\, da\, dt \tag{1.7}$$

$$= \int_0^T \int_0^{a_\dagger} f(a,t)\varphi(a,t) da\, dt,$$

$\forall \varphi \in \Phi$ (with Φ defined in Section 2.1). Passing to the limit in (1.7) we obtain

$$\int_0^T \int_0^{a_\dagger} [-D\varphi(a,t) + (\mu(a,t) + u^*(a,t))\varphi(a,t) - \beta(a,t)\varphi(0,t)] p^*(a,t) da\, dt$$

$$= \int_0^T \int_0^{a_\dagger} f(a,t)\varphi(a,t) da\, dt,$$

$\forall \varphi \in \Phi$ and by Theorem 2.1.4, (1.6) follows. Relations (1.5) and (1.6) imply the conclusion of the theorem.

The maximum principle

Concerning the optimality conditions, the following result holds:

THEOREM 3.1.2. *Suppose that in addition $f(a,t) > 0$ a.e. in Q_T. If u^* is an optimal control for problem* (**OH**) *and q is the solution of*

$$\begin{cases} Dq - \mu q = u^*(1+q) - \beta(a,t) q(0,t), & (a,t) \in Q_T, \\ q(a_\dagger, t) = 0, & t \in (0,T), \\ q(a,T) = 0, & a \in (0, a_\dagger), \end{cases} \tag{1.8}$$

then we have

$$u^*(a,t) = \begin{cases} \zeta_1(a,t), & \text{if } 1 + q(a,t) < 0, \\ \zeta_2(a,t), & \text{if } 1 + q(a,t) > 0. \end{cases} \tag{1.9}$$

Proof. Existence and uniqueness of the solution q to (1.8) follows by Lemma 2.1.5. Denote by $N_{\mathcal{U}}(u^*)$ the normal cone at $\mathcal{U}$ in u^* (in $L^\infty(Q_T)$).

For any $v \in L^\infty(Q_T)$, such that $u^* + \varepsilon v \in \mathcal{U}$ and for any $\varepsilon > 0$ small enough, we get

$$\int_0^T \int_0^{a_\dagger} u^* p^{u^*} da\, dt \geq \int_0^T \int_0^{a_\dagger} (u^* + \varepsilon v) p^{u^* + \varepsilon v} da\, dt$$

(because u^* is an optimal control for (**OH**)) and this implies

$$\int_0^T \int_0^{a_\dagger} u^* \frac{p^{u^*+\varepsilon v} - p^{u^*}}{\varepsilon} da\, dt + \int_0^T \int_0^{a_\dagger} v p^{u^*+\varepsilon v} da\, dt \leq 0. \tag{1.10}$$

The convergence $p^{u^*+\varepsilon v} \to p^{u^*}$ in $L^\infty(0,T; L^1(0,a_\dagger))$, as $\varepsilon \to 0^+$ follows by Theorem 2.1.2 (iii). In addition we need the next result:

LEMMA 3.1.3. *The following convergence holds:*

$$\frac{1}{\varepsilon}\left[p^{u^*+\varepsilon v} - p^{u^*}\right] \to z \quad \text{in } L^\infty(0,T; L^1(0,a_\dagger)) \text{ as } \varepsilon \to 0^+,$$

where z is the solution of

$$\begin{cases} Dz(a,t) + \mu(a,t) z(a,t) = -v p^{u^*} - u^* z & (a,t) \in Q_T, \\ z(0,t) = \displaystyle\int_0^{a_\dagger} \beta(a,t) z(a,t) da & t \in (0,T), \\ z(a,0) = 0 & a \in (0, a_\dagger). \end{cases} \tag{1.11}$$

Proof of Lemma 3.1.3. Existence and uniqueness of z, solution of (1.11) follows by Theorem 2.1.1. Denote by

$$w_\varepsilon(a,t) = \frac{1}{\varepsilon}\left[p^{u^*+\varepsilon v}(a,t) - p^{u^*}(a,t)\right] - z(a,t), \quad (a,t) \in Q_T.$$

It is obvious that w_ε is the solution of

$$\begin{cases} Dw(a,t) + \mu(a,t) w(a,t) = -u^* w - v\left[p^{u^*+\varepsilon v} - p^{u^*}\right] & (a,t) \in Q_T, \\ w(0,t) = \displaystyle\int_0^{a_\dagger} \beta(a,t) w(a,t) da & t \in (0,T), \\ w(a,0) = 0 & a \in (0, a_\dagger). \end{cases}$$

Since $p^{u^*+\varepsilon v} - p^{u^*} \to 0$ in $L^\infty(0,T;L^1(0,a_\dagger))$ as $\varepsilon \to 0^+$ we infer via Theorem 2.1.2 (iii) that $w_\varepsilon \to 0$ in $L^\infty(0,T;L^1(0,a_\dagger))$ as $\varepsilon \to 0^+$ and this concludes the proof of the lemma.

Proof of Theorem 3.1.2 - continued. Passing now to the limit in (1.10) we conclude

$$\int_0^T \int_0^{a_\dagger} (u^* z + v p^{u^*}) da\, dt \le 0. \tag{1.12}$$

Multiplying $(1.8)_1$ by z and integrating over $[0,T] \times [0,a_\dagger]$ we get:

$$\int_0^T \int_0^{a_\dagger} (Dq - \mu q) z\, da\, dt = \int_0^T \int_0^{a_\dagger} [u^*(1+q)z - \beta(a,t) q(0,t) z(a,t)] da\, dt$$

and after an easy calculation (involving (1.8) and (1.11)) we obtain that

$$-\int_0^T \int_0^{a_\dagger} q(Dz + \mu z) da\, dt = \int_0^T \int_0^{a_\dagger} u^*(1+q) z\, da\, dt. \tag{1.13}$$

Using now (1.11) and (1.13) we obtain:

$$\int_0^T \int_0^{a_\dagger} q v p^{u^*}\, da\, dt = \int_0^T \int_0^{a_\dagger} u^* z\, da\, dt, \tag{1.14}$$

and from (1.12) and (1.14) it follows that

$$\int_0^T \int_0^{a_\dagger} v(a,t)(1 + q(a,t)) p^{u^*}(a,t) da\, dt \le 0,$$

$\forall v \in L^\infty(Q_T)$, such that $u^* + \varepsilon v \in \mathcal{U}$, for any $\varepsilon > 0$ small enough. This is equivalent to

$$(1+q) p^{u^*} \in N_{\mathcal{U}}(u^*).$$

The last relation implies

$$u^*(a,t) = \begin{cases} \zeta_1(a,t), & \text{if } (1+q(a,t)) p^{u^*}(a,t) < 0, \\ \zeta_2(a,t), & \text{if } (1+q(a,t)) p^{u^*}(a,t) > 0. \end{cases}$$

Since $f(a,t) > 0$ a.e. in Q_T, we obtain via Theorem 2.1.2, that $p^{u^*}(a,t) > 0$ a.e. in Q_T and consequently (1.9) holds.

As a consequence of Theorem 3.1.2 we obtain that q (the solution of (1.8)) is a solution of

$$\begin{cases} Dq - (\mu + \zeta_1)q = \zeta_1 + (\zeta_2 - \zeta_1)(1+q)^+ & \\ \quad -\beta(a,t)q(0,t), & \text{in } Q_T, \\ q(a_\dagger, t) = 0, & \text{in } (0,T), \\ q(a,T) = 0, & \text{in } (0,a_\dagger). \end{cases} \tag{1.15}$$

In fact (1.15) has a unique solution. Indeed, if q_1 and q_2 are two solutions of (1.15), then $w := q_1 - q_2$ is a solution to

$$\begin{cases} Dw - (\mu + \zeta_1)w = (\zeta_2 - \zeta_1) \quad [(1+q_1)^+ - (1+q_2)^+] & \\ \quad -\beta(a,t)w(0,t), & (a,t) \in Q_T, \\ w(a_\dagger, t) = 0, & t \in (0,T), \\ w(a,T) = 0, & a \in (0,a_\dagger). \end{cases} \tag{1.16}$$

Since $|w|$ is absolutely continuous along almost every characteristic line (because w is), we deduce by (1.16) that $|w|$ is the solution to

$$\begin{cases} Dh - (\mu + \zeta_1)h = l(a,t) - \beta(a,t)h(0,t), & (a,t) \in Q_T, \\ h(a_\dagger, t) = 0, & t \in (0,T), \\ h(a,T) = 0, & a \in (0,a_\dagger), \end{cases}$$

where

$$\begin{aligned} l(a,t) &= (\zeta_2 - \zeta_1)[(1+q_1)^+ - (1+q_2)^+]sign(q_1 - q_2) \\ &\quad -\beta(a,t)w(0,t)sign(q_1 - q_2)(a,t) + \beta(a,t)|w(0,t)|. \end{aligned}$$

Since $l \in L^\infty(Q_T)$ and $l(a,t) \geq 0$ a.e. in Q_T, we deduce via Lemma 2.1.5 that $|w(a,t)| \leq 0$ a.e. in Q_T and consequently $q_1(a,t) = q_2(a,t)$ a.e. in Q_T.

The uniqueness of the optimal control is clarified by the following result:

THEOREM 3.1.4. *Let u^* be an optimal control for* (**OH**). *Suppose that $f(a,t) > 0$ a.e. in Q_T, and that the following assumption holds*

$$(\mathbf{A4}) \quad \begin{cases} \mu(a,t) > 0, \quad \text{a.e. in } Q_T, \\ \\ \text{and for almost any } t \in (0,T), \ \dfrac{\beta}{\mu}(\cdot,t) \\ \\ \text{is not a strictly positive constant on any subset} \\ \text{of positive measure .} \end{cases}$$

Under these additional conditions, equation $(1.8)_1$ *implies that the set*

$$B = \{(a,t) \in Q_T;\ q(a,t) = -1\}$$

has the Lebesgue measure zero (q *is the solution of* (1.8)) *and u^* is a bang-bang control* (which means that u^* takes only a finite number of values, almost everywhere).

In addition, the optimal control is unique.

Proof. Let q be the solution of (1.8). For almost any $(a,t) \in Int(B)$,

$$Dq(a,t) = 0$$

and consequently

$$\mu(a,t) = -\beta(a,t)q(0,t) \quad \text{a.e. } (a,t) \in B.$$

Assumption (**A4**) allows us to infer that $meas(B) = 0$ and so $q(a,t) \neq -1$ a.e. in Q_T. By (1.9) we deduce that u^* is a bang-bang control. Since q is also the unique solution of (1.15) we may conclude via (1.9) the uniqueness of the optimal control.

REMARK 3.1.1. *If the hypotheses in Theorem 3.1.4 hold, then we conclude that the adjoint state q (which is also the solution of* (1.15)) *does not depend on f and on p_0.*

Denote by u^* this optimal control.

The last remark allows us to formulate the following result for the general case $f(a,t) \geq 0$ a.e. in Q_T:

THEOREM 3.1.5. *If* (**A1**) – (**A4**) *hold, then the control u^* is optimal also for problem* (**OH**) *corresponding to the nonnegative flow f.*

Proof. Consider $f_n \in L^\infty(0,T;L^1(0,a_\dagger))$, such that $f_n(a,t) > 0$ a.e. in Q_T and $f_n \to f$ in $L^\infty(0,T;L^1(0,a_\dagger))$ (as $n \to +\infty$). Denote by $\Psi_n(u)$ the cost function corresponding to $f := f_n$ in problem (**OH**) and by p_n^u the solution to (1.1) corresponding to u and $f := f_n$.

Theorem 2.1.2 allows us to conclude that for any $u \in \mathcal{U}$, $p_n^u \to p^u$ in $L^\infty(0,T;L^1(0,a_\dagger))$, as $n \to +\infty$. This implies

$$\Psi_n(u) \to \Psi(u) \quad \text{as } n \to +\infty. \tag{1.17}$$

Since for any $u \in \mathcal{U}$ we have $\Psi_n(u^*) \geq \Psi_n(u)$, and using (1.17), we conclude that u^* is an optimal control for (**OH**).

REMARK 3.1.2. *If* (**A1**) – (**A4**) *hold, then* (**OH**) *has a unique optimal control* u^*. *In order to find* u^* *we find first the solution* q *of* (1.15). *The optimal control is given now by* (1.9).

We propose to the reader the study of the following slightly more general problem:

$$(\mathbf{OH_1}) \qquad \textit{Maximize} \int_0^T \int_0^{a_\dagger} u(a,t)g(a)p^u(a,t)da\, dt,$$

subject to $u \in \mathcal{U}$ (where p^u is the solution of (1.1)).

The function g satisfies

$$g \in C^1([0,a_\dagger]) \;,\; g(a) > 0 \quad \text{for any } a \in (0,a_\dagger).$$

Here $g(a)$ signifies the weight of an individual of age a and

$$\int_0^T \int_0^{a_\dagger} u(a,t)g(a)p^u(a,t)da\, dt$$

is the total weight of the harvested population.

3.2. Optimal harvesting for a nonlinear model

Here we investigate an optimal harvesting problem for a nonlinear age dependent population dynamics. We establish the existence of the optimal effort and the convergence of a certain fractional step scheme. For some approximating problems we obtain the optimal controllers in feedback form via the dynamic programming method.

Consider the following problem:

$$(\mathbf{OH}) \qquad \textit{Maximize} \int_0^T \int_0^{a_\dagger} u(t)g(a)p^u(a,t)da\, dt,$$

subject to $u \in \mathcal{U}$ and p^u is the solution of

$$\begin{cases} (Dp + \mu_0 p)(a,t) + \mathcal{M}(P(t))p(a,t) = -u(t)p(a,t), & (a,t) \in Q_T, \\ p(0,t) = \displaystyle\int_0^{a_\dagger} \beta(a,t)p(a,t)da, & t \in (0,T), \\ p(a,0) = p_0(a) & a \in (0,a_\dagger), \\ P(t) = \displaystyle\int_0^{a_\dagger} p(a,t)da & t \in (0,T). \end{cases} \tag{2.1}$$

$g(a)$ is a weight, which represents the cost of an individual of age a. So, the integral $\int_0^T \int_0^{a_\dagger} u(t)g(a)p^u(a,t)da\, dt$ gives the total cost of the harvested population. Here

$$\mathcal{U} = \{v \in L^\infty(0,T);\ 0 \le v(t) \le L \text{ a.e. in } (0,T)\}$$

and we suppose the following assumptions hold (see Section 2.2):

(As0) $$g \in C^1([0,a_\dagger])\ ,\ g(a) > 0 \text{ a.e. in } (0,a_\dagger);$$

(As1) $$\begin{cases} \beta \in L^\infty(Q_T)\ ,\ \beta(a,t) \ge 0 \quad \text{a.e. in } Q_T, \\ \beta(a,t) > 0 \quad \text{a.e. in } (\alpha_1,\alpha_2) \times (0,T), \end{cases}$$

where $0 \le \alpha_1 < \alpha_2 \le a_\dagger$;

(As2) $$\begin{cases} \mu_0 \in L^1_{loc}([0,a_\dagger) \times [0,T])\ ,\ \mu_0(a,t) \ge 0 \quad \text{a.e. in } Q_T, \\ \displaystyle\int_0^{a_\dagger} \mu_0(a, t - a_\dagger + a)da = +\infty, \\ \text{where } \mu_0(a,t) \text{ is extended by zero on } (0,a_\dagger) \times (-\infty,0); \end{cases}$$

(As3) $\mathcal{M} : [0,+\infty) \to [0,+\infty)$ is a continuously differentiable and nondecreasing function,

and the initial density p_0 satisfies

(As4) $$p_0 \in L^1(0,A)\ ,\ p_0(a) > 0 \quad \text{a.e. in } (0,a_\dagger).$$

We have proved in Section 2.2 that (2.1) belongs to the class of "separable" models considered in [27], [10], i.e., system (2.1) has a unique solution p^u and

$$p^u(a,t) = y^u(t)\tilde{p}(a,t), \tag{2.2}$$

where $\tilde{p}$ is the solution of the "free" problem

$$\begin{cases} D\tilde{p} + \mu_0(a,t)\tilde{p} = 0, & (a,t) \in Q_T, \\ \tilde{p}(0,t) = \int_0^{a_\dagger} \beta(a,t)\tilde{p}(a,t)da, & t \in (0,T), \\ \tilde{p}(a,0) = p_0(a), & a \in (0,a_\dagger), \end{cases}$$

and y^u is the Carathéodory solution of

$$\begin{cases} y'(t) + \mathcal{M}(P_0(t)y(t))y(t) + u(t)y(t) = 0, & t \in (0,T), \\ y(0) = 1 \ . \end{cases}$$

Since β satisfies **(As1)** and p_0 satisfies **(As4)**, we immediately conclude that $\tilde{p}$ is strictly positive almost everywhere. It follows that p^u is also strictly positive almost everywhere. Problem **(OH)** can be reduced to the following one on y^u:

$$(\mathbf{OH_0}) \qquad \textit{Maximize} \int_0^T m(t)u(t)y^u(t)dt,$$

subject to $u \in \mathcal{U}$, where $m(t) = \int_0^{a_\dagger} g(a)\tilde{p}(a,t)da$, $t \in [0,T]$ is known (the assumption on g implies that $m \in L^\infty(0,T)$, $m(t) > 0$ a.e. in $(0,T)$).

In conclusion, the study of problem **(OH)** can be reduced to the study of $(\mathbf{OH_0})$ via the representation formula (2.2). Indeed,

$$\int_0^T \int_0^{a_\dagger} u(t)g(a)p^u(a,t)da\ dt = \int_0^T m(t)u(t)y^u(t)dt, \tag{2.3}$$

and so the optimal controls for **(OH)** and $(\mathbf{OH_0})$ are the same and via (2.3) we get that the optimal values of the cost functionals are the same.

Thus our attention will be focused on $(\mathbf{OH_0})$, and any result for this problem can be easily translated into a result for the original problem.

We note that $(\mathbf{OH_0})$ actually depends on the initial datum p_0 via the term P_0.

We shall start with a slightly more general problem:

$$(\mathbf{OH_{00}}) \qquad \textit{Maximize} \int_0^T m(t)u(t)h^u(t)dt,$$

subject to $u \in \mathcal{U}$ and h^u is the Carathéodory solution of

$$\begin{cases} h'(t) + \mathcal{M}(P_0(t)h(t))h(t) + u(t)h(t) = 0, & t \in (0,T), \\ h(0) = h_0 \in (0,+\infty). \end{cases} \tag{2.4}$$

The first result in this section concerns the existence of an optimal control for $(\mathbf{OH_{00}})$.

THEOREM 3.2.1. *There exists at least one optimal control for* $(\mathbf{OH_{00}})$.

We shall use as the main ingredient the following

LEMMA 3.2.2. *If* $\{u_n\} \subset \mathcal{U}$ *satisfies* $u_n \to u$ *weakly in* $L^2(0,T)$, *then*

$$h^{u_n} \to h^u \quad \text{in } L^2(0,T).$$

Proof of the Lemma 3.2.2. Since h^{u_n} is the Carathéodory solution of (2.4) with $u := u_n$, we infer that

$$h^{u_n}(t) = \exp\Big\{ - \int_0^t (u_n(s) + \mathcal{M}(P_0(s)h^{u_n}(s)))ds \Big\} \cdot h_0,$$

for any $t \in [0,T]$ and this implies

$$0 \le h^{u_n}(t) \le h_0, \quad \text{for any } t \in [0,T].$$

The last relation allows us to conclude that the sequence $\{v_n\}$ given by

$$v_n(t) = \mathcal{M}(P_0(t)h^{u_n}(t)), \quad \text{a.e. } t \in (0,T),$$

satisfies

$$0 \le v_n(t) \le M, \quad \text{a.e. } t \in (0,T),$$

where $M \in (0,+\infty)$ is a constant. On a subsequence (also denoted by $\{v_n\}$) we have

$$v_n \to v \quad \text{weakly in } L^2(0,T). \tag{2.5}$$

Since $u_n \to u$ weakly in $L^2(0,T)$, we conclude that

$$h^{u_n} \to \tilde{h}^v \quad \text{in } L^2(0,T),$$

where $\tilde{h}^v$ is the Carathéodory solution to

$$\begin{cases} h'(t) + v(t)h(t) = -u(t)h(t), & t \in (0,T), \\ h(0) = h_0 . \end{cases}$$

By (2.5) we get

$$v_n(\cdot) = \mathcal{M}(P_0(\cdot)h^{u_n}(\cdot)) \to \mathcal{M}(P_0(\cdot)\tilde{h}^v(\cdot)) \quad \text{in } L^2(0,T)$$

as $n \to +\infty$. So, the conclusion is that $v(\cdot) = \mathcal{M}(P_0(\cdot)\tilde{h}^v(\cdot))$ and $\tilde{h}^v \equiv h^u$.

Proof of Theorem 3.2.1. Consider now

$$d = \sup_{u \in \mathcal{U}} \int_0^T m(t)u(t)h^u(t)dt .$$

It is obvious that $d \in [0, +\infty)$. Let $u_n \in \mathcal{U}$ be such that

$$d - \frac{1}{n} < \int_0^T m(t)u_n(t)h^{u_n}(t)dt \leq d .$$

As a consequence we obtain that on a subsequence (also denoted by $\{u_n\}$) we have

$$u_n \to u^* \quad \text{weakly in } L^2(0,T) . \tag{2.6}$$

By Lemma 3.2.2 we obtain that

$$h^{u_n} \to h^{u^*} \quad \text{in } L^2(0,T),$$

and consequently

$$mh^{u_n} \to mh^{u^*} \quad \text{in } L^2(0,T)$$

(because $m \in L^\infty(0,T)$). Therefore

$$\int_0^T m(t)u_n(t)h^{u_n}(t)dt \to \int_0^T m(t)u^*(t)h^{u^*}(t)dt \tag{2.7}$$

and (by (2.6) and (2.7))

$$d = \int_0^T m(t)u^*(t)h^{u^*}(t)dt .$$

We may infer now that (u^*, h^{u^*}) is an optimal pair for $(\mathbf{OH_{00}})$.

AN APPROXIMATING SCHEME

We shall prove now that problem $(\mathbf{OH_{00}})$ can be approximated (in a certain sense) for $\varepsilon \to 0^+$ by the following sequence of optimal control problems:

$$(\mathbf{OH_{00\varepsilon}}) \qquad \text{Maximize} \int_0^T m(t)u(t)h_\varepsilon^u(t)dt,$$

subject to $u \in \mathcal{U}$ and h_ε^u is the Carathéodory solution of

$$\begin{cases} h'(t) + \gamma(t)h(t) = -u(t)h(t), \quad t \in (i\varepsilon, (i+1)\varepsilon), \\ h(i\varepsilon+) = \theta((i+1)\varepsilon-; i\varepsilon, h(i\varepsilon-)), \quad i = 0, 1, ..., N-1, \ \varepsilon = \frac{T}{N}, \\ h(0-) = h_0 \in (0, +\infty), \end{cases} \tag{2.8}$$

where $\theta(t; i\varepsilon, x)$ is the Carathéodory solution of

$$\begin{cases} \theta'(t) + \mathcal{M}(P_0(t)\theta(t))\theta(t) = \gamma(t)h(t), \quad t \in (i\varepsilon, (i+1)\varepsilon), \\ \theta(i\varepsilon+) = x. \end{cases} \tag{2.9}$$

Here we choose γ such that

$$(\mathbf{As5}) \qquad \gamma \in C([0, T]).$$

For other results concerning some fractional step schemes we refer to [5], [14], [15].

Using an analogous argument as above it is possible to prove that $(\mathbf{OH_{00\varepsilon}})$ has at least one optimal pair.

First of all we need to establish the following technical result:

LEMMA 3.2.3. *If $u_\varepsilon \to u$ weakly in $L^2(0,T)$ for $\varepsilon \to 0^+$ $(u_\varepsilon \in \mathcal{U})$, then*

$$h_\varepsilon^{u_\varepsilon} \to h^u \quad \text{in } BV([0,T]),$$

as $\varepsilon \to 0^+$.

Proof. By (2.8) and (2.9) we obtain that for any $i\varepsilon \le t < (i+1)\varepsilon$:

$$\begin{aligned} h_\varepsilon^{u_\varepsilon}(t) &= \exp\Big\{-\int_{i\varepsilon}^t (u_\varepsilon + \gamma)(s)ds\Big\}\theta((i+1)\varepsilon-; i\varepsilon, h_\varepsilon^{u_\varepsilon}(i\varepsilon-)) \\ &= \exp\Big\{-\int_{i\varepsilon}^t (u_\varepsilon + \gamma)(s)ds + \int_{i\varepsilon}^{(i+1)\varepsilon} (\gamma - v_\varepsilon)(s)ds\Big\}h_\varepsilon^{u_\varepsilon}(i\varepsilon-), \end{aligned} \tag{2.10}$$

where

$$v_\varepsilon(s) = \mathcal{M}(P_0(s)\theta(s; i\varepsilon, h_\varepsilon^{u_\varepsilon}(i\varepsilon-))), \quad \text{a.e. } s \in (0, T).$$

The sequence $\{v_\varepsilon\}$ is bounded in $L^\infty(0,T)$ and nonnegative. It is obvious now that (2.10) implies the existence of a constant $M \in (0,+\infty)$ such that

$$|h_\varepsilon^{u_\varepsilon}(t)| \le M, \quad \text{for all } t \in [0,T].$$

Relation $(2.8)_1$ implies that

$$\begin{aligned}\int_0^T |(h_\varepsilon^{u_\varepsilon})'(t)|dt &= \int_0^T |\gamma(t)+u(t)|\cdot|h_\varepsilon^{u_\varepsilon}(t)|dt \\ &\le MT(L+\|\gamma\|_{C([0,T])})\end{aligned} \tag{2.11}$$

and

$$\begin{aligned}|h_\varepsilon^{u_\varepsilon}(i\varepsilon+)-h_\varepsilon^{u_\varepsilon}(i\varepsilon-)| &= \left|e^{\int_{i\varepsilon}^{(i+1)\varepsilon}(\gamma-v_\varepsilon)(s)ds}-1\right|\cdot|h_\varepsilon^{u_\varepsilon}(i\varepsilon-)| \\ &\le M\cdot\left|\int_{i\varepsilon}^{(i+1)\varepsilon}(\gamma(s)-v_\varepsilon(s))ds\right|.\end{aligned} \tag{2.12}$$

By (2.12) we see that

$$\sum_{i=0}^{N}|h_\varepsilon^{u_\varepsilon}(i\varepsilon+)-h_\varepsilon^{u_\varepsilon}(i\varepsilon-)| \le M_1 \tag{2.13}$$

(where $M_1 \in (0,+\infty)$ is a constant) and now (2.11) and (2.13) imply that the total variation of $h_\varepsilon^{u_\varepsilon}$ on $[0,T]$ is bounded by the same constant (independent of ε). We conclude via Helly's theorem that on a subsequence (also denoted by $\{h_\varepsilon^{u_\varepsilon}\}$) we have

$$h_\varepsilon^{u_\varepsilon} \to h \quad \text{in } BV([0,T])$$

and consequently

$$h_\varepsilon^{u_\varepsilon}(t) \to h(t) \quad \text{for all } t \in [0,T]$$

and

$$h_\varepsilon^{u_\varepsilon} \to h \quad \text{in } L^2(0,T).$$

Let us prove that $h^u \equiv h$. Indeed, from (2.10) we obtain that for any $i\varepsilon \le t < (i+1)\varepsilon$:

$$h_\varepsilon^{u_\varepsilon}(t) = \exp\Big\{-\int_0^t u_\varepsilon(s)ds - \int_0^{(i+1)\varepsilon} v_\varepsilon(s)ds - \int_{i\varepsilon}^t \gamma(s)ds\Big\}\cdot h_0. \tag{2.14}$$

Since

$$\exp\Big\{-\int_0^t u_\varepsilon(s)ds\Big\} \to \exp\Big\{-\int_0^t u(s)ds\Big\} \quad \text{in } C([0,T]),$$

$$\exp\Big\{\int_{i\varepsilon}^{t}\gamma(s)ds\Big\}\to 1\quad\text{in } BV([0,T])$$

and

$$\exp\Big\{-\int_{0}^{(i+1)\varepsilon}v_\varepsilon(s)ds\Big\}\to\exp\Big\{-\int_{0}^{(i+1)\varepsilon}\tilde v(s)ds\Big\}\quad\text{in } BV([0,T])$$

(where, on a subsequence also denoted by $\{v_\varepsilon\}$ we have $v_\varepsilon\to\tilde v$ in $L^2(0,T)$), we obtain by passing to the limit in (2.14) that h is the Carathéodory solution of

$$\begin{cases} h'(t)+\tilde v(t)h(t)=-u(t)h(t), & t\in(0,T),\\ h(0)=h_0. \end{cases}\tag{2.15}$$

The convergence

$$v_\varepsilon(t)=\mathcal{M}(P_0(t)h_\varepsilon^{u_\varepsilon}(t))\to\mathcal{M}(P_0(t)h(t)),\quad\text{a.e. in }(0,T),$$

for $\varepsilon\to 0^+$ implies

$$\tilde v(t)=\mathcal{M}(P_0(t)h(t)),\quad\text{a.e. in }(0,T)$$

and by (2.15) we get that $h^u\equiv h$.

The proof of the Lemma 3.2.3 is now complete.

Consider $\varphi,\ \varphi_\varepsilon:\mathcal{U}\to[0,+\infty)$ defined by

$$\varphi(u)=\int_0^T m(t)u(t)h^u(t)dt$$

and

$$\varphi_\varepsilon(u)=\int_0^T m(t)u(t)h_\varepsilon^u(t)dt$$

and u_ε^* an optimal control for $(\mathbf{OH}_{00\varepsilon})$.

We conclude this subsection with the following main result:

THEOREM 3.2.4. *If u^* is a weak limit point of $\{u_\varepsilon^*\}$ in $L^2(0,T)$, then u^* is an optimal control for $(\mathbf{OH}_{00})$ and in addition*

$$\lim_{\varepsilon\to 0^+}\varphi(u_\varepsilon^*)=\varphi(u^*)\tag{2.16}$$

and

$$\lim_{\varepsilon\to 0^+}\varphi_\varepsilon(u_\varepsilon^*)=\varphi(u^*).\tag{2.17}$$

Proof. Noting that

$$\varphi_\varepsilon(u_\varepsilon^*) = \int_0^T m(t)u_\varepsilon^*(t)h_\varepsilon^{u_\varepsilon^*}(t)dt \geq \int_0^T m(t)u(t)h_\varepsilon^u(t)dt,$$

for any $u \in \mathcal{U}$ and using the previous lemma we conclude that

$$\int_0^T m(t)u^*(t)h^{u^*}(t)dt \geq \int_0^T m(t)u(t)h^u(t)dt,$$

for any $u \in \mathcal{U}$. This implies that u^* is an optimal control for $(\mathbf{OH_{00}})$.

Now, since the following convergences hold,

$$u_\varepsilon^* \to u^* \quad \text{weakly in } L^2(0,T)$$

and

$$h_\varepsilon^{u_\varepsilon^*} \to h^{u^*} \quad \text{in } L^2(0,T),$$

we may conclude that (2.17) holds.

Using now the convergence

$$h^{u_\varepsilon^*} \to h^{u^*} \quad \text{in } L^2(0,T)$$

(see Lemma 3.2.3) we obtain relation (2.16).

THE DYNAMIC PROGRAMMING METHOD

Consider the following problems:

$$(\mathbf{OH_{00\varepsilon,t}}) \qquad \text{Maximize} \int_t^T m(s)u(s)h_\varepsilon^u(s;t,x)ds,$$

subject to $u \in \mathcal{U}_t = \{v \in L^\infty(t,T);\ 0 \leq v(s) \leq L \text{ a.e. }\}$ and h_ε^u the Carathéodory solution of

$$\begin{cases} h'(t) + \gamma(t)h(t) = -u(t)h(t), & s \in (t,T] \setminus \{i\varepsilon\}_{i=0}^N, \\ h(i\varepsilon+) = \theta((i+1)\varepsilon-; i\varepsilon, h_\varepsilon^u((i+1)\varepsilon-;t,x)), \\ h(t-) = x. \end{cases} \tag{2.18}$$

Here $(t,x) \in [0,T] \times (0,+\infty)$ and $\theta(s;t,x)$ is the Carathéodory solution of

$$\begin{cases} \theta'(s) + \mathcal{M}(P_0(s)\theta(s))\theta(s) = \gamma(s)h(s), & s \in (i\varepsilon,(i+1)\varepsilon) \cap (t,T], \\ \theta(i\varepsilon+) = x. \end{cases}$$

As in Theorem 3.2.1, there follows the existence of an optimal pair $(u_\varepsilon(\cdot;t,x), h_\varepsilon(\cdot;t,x))$ for $(\mathbf{OH}_{\mathbf{00}\varepsilon,\mathbf{t}})$.
Suppose that in addition the following assumptions hold:

$$(\mathbf{As6}) \qquad m \in W^{1,1}(0,T)$$

(which is fullfiled under certain additional assumptions on p_0; for example if $p_0 \in W^{1,1}(0,a_\dagger)$ and $p_0(0) = \int_0^{a_\dagger} \beta(a)p_0(a)da$ - see Theorem 2.1.3) and

$$(\mathbf{As7}) \qquad \gamma - \frac{m'}{m} \neq 0$$

almost everywhere (γ is choosen in order to satisfy this condition and **(As5)**).

Denote by

$$\psi^\varepsilon(t,x) = Max_{u\in\mathcal{U}_t} \int_t^T m(s)u(s)h_\varepsilon^u(s;t,x)ds.$$

It is obvious that for all $0 \le t \le s \le T$ and $x \in (0,+\infty)$, $\psi^\varepsilon(t,x)$ is the optimal value for the following problem:

$$(\mathbf{OH}_{\mathbf{00}\varepsilon,\mathbf{t},\mathbf{s}}) \qquad \text{Maximize} \int_t^s m(\tau)u(\tau)h_\varepsilon^u(\tau;t,x)d\tau + \psi^\varepsilon(s, h_\varepsilon^u(s;t,x)),$$

subject to $u \in L^\infty(t,s)$, $0 \le u(\tau) \le L$ a.e. (and h_ε^u is the Carathéodory solution of (2.18) on [t,s]) and $(u_\varepsilon|_{[t,s]}, h_\varepsilon|_{[t,s]})$ is an optimal pair for this last maximization problem.

It is also clear that

$$\psi^\varepsilon(i\varepsilon-,x) = \psi^\varepsilon(i\varepsilon+, \theta((i+1)\varepsilon-;i\varepsilon,x)), \quad i = 0,1,...,N-1$$

and that
ψ^ε is locally Lipschitz on $(i\varepsilon,(i+1)\varepsilon) \times (0,+\infty)$, $i = 0,1,...,N-1$ (e.g., [16]).

By virtue of Rademacher's Theorem, the function ψ^ε is almost everywhere derivable on $[0,T] \times (0,+\infty)$.

The optimality conditions for $(\mathbf{OH}_{\mathbf{00}\varepsilon,\mathbf{t},(\mathbf{i+1})\varepsilon})$ $(i\varepsilon < t < (i+1)\varepsilon\ )$ are given by

THEOREM 3.2.5. *There exists an absolutely continuous function q which is a Carathéodory solution to*

$$\begin{cases} q' - \gamma(s)q + \dfrac{m'(s)}{m(s)}q = u_\varepsilon(s;t,x)(1+q(s)), \quad s \in (t,(i+1)\varepsilon), \\ \\ q((i+1)\varepsilon-) \in -\dfrac{1}{m((i+1)\varepsilon)}\partial_x\psi^\varepsilon((i+1)\varepsilon-, h_\varepsilon((i+1)\varepsilon-;t,x)), \end{cases} \tag{2.19}$$

and

$$u_\varepsilon(s;t,x) = \begin{cases} 0, & \text{if } 1+q(s) < 0, \\ \\ L, & \text{if } 1+q(s) > 0. \end{cases} \tag{2.20}$$

Here $\partial_x\psi^\varepsilon(\cdot,\cdot)$ denotes the Clarke's generalized gradient with respect to the second variable (see Appendix 1).

Proof. The proof is standard. We shall sketch it only under the additional assumption $\psi^\varepsilon((i+1)\varepsilon-,\cdot) \in C^1(0,+\infty)$. In this case $\partial_x\psi^\varepsilon = \dfrac{\partial\psi^\varepsilon}{\partial x}$.

Since $(u_\varepsilon(\cdot;t,x), h_\varepsilon(\cdot;t,x))$ is an optimal pair, then

$$\int_t^{(i+1)\varepsilon} m(s)u_\varepsilon(s)h_\varepsilon(s;t,x)ds + \psi^\varepsilon((i+1)\varepsilon-, h_\varepsilon((i+1)\varepsilon-;t,x))$$

$$\geq \int_t^{(i+1)\varepsilon} m(s)(u_\varepsilon+\eta v)(s)h_\varepsilon^{u_\varepsilon+\eta v}(s;t,x)ds$$
$$+\psi^\varepsilon((i+1)\varepsilon-, h_\varepsilon^{u_\varepsilon+\eta v}((i+1)\varepsilon-;t,x)),$$

for any $v \in L^\infty(t,T)$ such that $u_\varepsilon + \eta v \in \mathcal{U}_t$ for any $\eta > 0$ small enough, and consequently

$$\int_t^{(i+1)\varepsilon} mu_\varepsilon \frac{h_\varepsilon^{u_\varepsilon+\eta v} - h_\varepsilon}{\eta}(s;t,x)ds + \int_t^{(i+1)\varepsilon} mvh_\varepsilon^{u_\varepsilon+\eta v}(s;t,x)ds$$

$$+\frac{1}{\eta}\left[\psi^\varepsilon((i+1)\varepsilon, h_\varepsilon^{u_\varepsilon+\eta v}((i+1)\varepsilon-;t,x)) - \psi^\varepsilon((i+1)\varepsilon, h_\varepsilon((i+1)\varepsilon-;t,x))\right] \leq 0.$$

Passing to the limit ($\eta \to 0^+$) we obtain

$$\int_t^{(i+1)\varepsilon} m(s)(u_\varepsilon z + vh_\varepsilon)(s;t,x)ds$$

$$+\frac{\partial}{\partial x}\psi^\varepsilon((i+1)\varepsilon-, h_\varepsilon((i+1)\varepsilon-;t,x)) \leq 0, \tag{2.21}$$

for any v as above, where z is the Carathéodory solution of

$$\begin{cases} z' + \gamma(s)z = -u_\varepsilon(s)z(s) - v(s)h_\varepsilon(s;t,x), & s \in (t,(i+1)\varepsilon], \\ z(t) = 0. \end{cases} \tag{2.22}$$

Let q be the solution of (2.19). Multiplying $(2.19)_1$ by $m(s)z(s)$ and integrating on $[t,(i+1)\varepsilon]$ we obtain

$$\int_t^{(i+1)\varepsilon} q'(s)m(s)z(s)ds - \int_t^{(i+1)\varepsilon} \gamma(s)q(s)m(s)z(s)ds$$

$$+\int_t^{(i+1)\varepsilon} m'(s)q(s)z(s)ds = \int_t^{(i+1)\varepsilon} u_\varepsilon(s)(1+q(s))m(s)z(s)ds.$$

After an easy calculation involving $(2.19)_2$ and (2.22) we obtain

$$-\frac{\partial}{\partial x}\psi^\varepsilon((i+1)\varepsilon, h_\varepsilon((i+1)\varepsilon-;t,x))z((i+1)\varepsilon)$$

$$+\int_t^{(i+1)\varepsilon} q(s)m(s)v(s)h_\varepsilon(s;t,x)ds = \int_t^{(i+1)\varepsilon} u_\varepsilon(s)m(s)z(s)ds,$$

for any $v \in L^\infty(t,T)$ such that $u_\varepsilon + \eta v \in \mathcal{U}_t$ for any $\eta > 0$ small enough. The last relation and (2.21) imply

$$\int_t^{(i+1)\varepsilon} m(s)v(s)(1+q(s))h_\varepsilon(s)ds \le 0,$$

for any $v \in L^\infty(t,T)$ such that $u_\varepsilon + \eta v \in \mathcal{U}_t$ for any $\eta > 0$ small enough, i.e.,

$$u_\varepsilon(s;t,x) = \begin{cases} 0, & \text{if } m(s)(1+q(s))h_\varepsilon(s;t,x) < 0, \\ L, & \text{if } m(s)(1+q(s))h_\varepsilon(s;t,x) > 0. \end{cases}$$

Since $m(s) > 0$ and $h_\varepsilon(s;t,x) > 0$ a.e. in $(0,T)$ we may conclude that (2.20) holds.

REMARK 3.2.1. *By Theorem 3.2.5 and hypothesis* (**As7**) *we conclude that the optimal control* u_ε *is of bang-bang type.*

In the same manner as in Theorem 3.2.5 it is possible to prove

LEMMA 3.2.6.

$$q(s) = -\frac{1}{m(s)} \cdot \partial_x \psi^\varepsilon(s, h_\varepsilon(s;t,x)), \quad \text{a.e. } s \in (t,(i+1)\varepsilon)$$

(*where q is given by Theorem 3.2.5*).

For any $i\varepsilon \le t \le s \le (i+1)\varepsilon$ we have

$$\psi^\varepsilon(t,x) - \psi^\varepsilon(s, h_\varepsilon(s;t,x)) = \int_t^s m(\tau)u_\varepsilon(\tau;t,x)h_\varepsilon(\tau;t,x)d\tau$$

and consequently

$$-\frac{\partial\psi^\varepsilon}{\partial t}(t,x) - \frac{\partial\psi^\varepsilon}{\partial x}(t,x)h'_\varepsilon(t;t,x) = m(t)u_\varepsilon(t)h_\varepsilon(t;t,x), \tag{2.23}$$

a.e. in $(0,T)\times(0,+\infty)$. Relation (2.23) and (2.20) imply

$$-\frac{\partial\psi^\varepsilon}{\partial t}(t,x) + \frac{\partial\psi^\varepsilon}{\partial x}(t,x)(\gamma(t) + L\cdot H(1+q(t)))x = m(t)L\cdot H(1+q(t))x,$$

a.e. in $(0,T)\times(0,+\infty)$ (here H is the Heaviside function). We may conclude now this section by the following result:

THEOREM 3.2.7. *The optimal value function ψ^ε satisfies the following Hamilton-Jacobi equation:*

$$\frac{\partial\psi^\varepsilon}{\partial t}(t,x) - \gamma(t)\frac{\partial\psi^\varepsilon}{\partial x}(t,x) + L\cdot\left(m(t) - \frac{\partial\psi^\varepsilon}{\partial x}(t,x)\right)^+ x = 0,$$

a.e. $(t,x)\in(0,T)\times(0,+\infty)$,

$$\psi^\varepsilon(i\varepsilon-,x) = \psi^\varepsilon(i\varepsilon+, \theta((i+1)\varepsilon-; i\varepsilon, x)), \quad i = 0,1,...,N-1,$$

$$\psi^\varepsilon(T,x) = 0, \ \forall x\in(0,+\infty),$$

and

$$u_\varepsilon(t;0,x) \in L\cdot H(m(t) - \frac{\partial\psi^\varepsilon}{\partial x}(t, h_\varepsilon(t;0,x)).$$

In fact we have found the optimal control in a feedback form.

Numerical algorithms for the approximation of the optimal controller u^* for (**OH**) can be developed from Theorems 3.2.1 and Theorem 3.2.7. For numerical results and numerical tests concerning some Hamilton-Jacobi equations we refer to [12].

REMARK 3.2.2. *If we choose γ to satisfy in addition*

$$\gamma(t) > \frac{m'(t)}{m(t)},$$

a.e. $t \in (0,T)$, then in any interval $(i\varepsilon,(i+1)\varepsilon)$ ($i \in \{0,\ 1, ...,\ N-1\}$), the function q has at most one point where it takes the value -1.

Indeed, for any $\tau \in (i\varepsilon,(i+1)\varepsilon)$ such that $q(\tau) = -1$ we have $q'(t) < 0$, for t in a neighbourhood of τ. This implies that there is at most one point with this property in the interval $\tau \in (i\varepsilon,(i+1)\varepsilon)$.

As a consequence, we have that $1 + q$ has at most one zero on every interval $(i\varepsilon,(i+1)\varepsilon)$ and consequently u_ε has the form

$$u_\varepsilon(t) = \begin{cases} L, & \text{if } t \in [i\varepsilon,\tau], \\ 0, & \text{if } t \in [\tau,(i+1)\varepsilon], \end{cases} \tag{2.24}$$

where τ is a certain point in $[i\varepsilon,(i+1)\varepsilon]$ (u_ε has at most one switching point in $[i\varepsilon,(i+1)\varepsilon]$).

CONCLUSION. The fractional step scheme we have used allows us to conclude that there is a sequence of bang-bang controllers with the structure as in (2.24), such that

$$\lim_{\varepsilon \to 0+} \varphi(u_\varepsilon^*) = \varphi(u^*)$$

(the optimal harvest is "approximated" by the harvest corresponding to the effort u_ε).

Relation (2.24) allows us to approximate the optimal harvest ($\varphi(u^*)$).

3.3. Optimal harvesting for a linear periodic population dynamics

Here we investigate optimal harvesting for periodic age-dependent population dynamics. Namely we shall study the following problem:

$$\textbf{(P)} \qquad Maximize \int_0^T \int_0^{a_\dagger} u(a,t)p^u(a,t)da\, dt,$$

over all $u \in \mathcal{U}$, where p^u is the solution of

$$\begin{cases} Dp + \mu(a,t)p = f(a,t) - u(a,t)p, & (a,t) \in Q, \\ p(0,t) = \int_0^{a_\dagger} \beta(a,t)p(a,t)da, & t \in \mathbf{R}, \\ p(a,t) = p(a,t+T), & (a,t) \in Q\ , \end{cases} \tag{3.1}$$

where $Q = [0, a_\dagger) \times \mathbf{R}$ ($a_\dagger$, $T \in (0,+\infty)$). The set of controls is

$$\mathcal{U} = \{h \in L_T^\infty(Q);\ \zeta_1(a,t) \le h(a,t) \le \zeta_2(a,t) \text{ a.e. in } Q\}\ ,$$

where ζ_1, $\zeta_2 \in L_T^\infty(Q)$, $0 \le \zeta_1(a,t) \le \zeta_2(a,t)$ a.e. in Q (the spaces $L_T^\infty(Q)$ and $L_T^\infty(\mathbf{R})$ were defined in Section 2.4).

The hypotheses we will use are those in Section 2.4:

(A1) $$\beta \in L_T^\infty(Q), \quad \beta(a,t) \geq 0 \quad \text{a.e. in } Q,$$

(A2) $$\begin{cases} \mu(a,t) = \mu_0(a) + \overline{\mu}(a,t) \quad \text{a.e. in } Q, \\ \text{where} \\ \mu_0 \in L^1_{loc}([0,a_\dagger)) \ , \ \mu_0(a) \geq 0 \quad \text{a.e. in } [0,a_\dagger), \\ \int_0^{a_\dagger} \mu_0(a)da = +\infty, \\ \overline{\mu} \in L_T^\infty(Q) \ , \ \overline{\mu}(a,t) \geq 0 \quad \text{a.e. in } Q, \end{cases}$$

(A3) $$f \in L_T^\infty(Q),$$

and in addition

(A4) $$f(a,t) \geq 0 \quad \text{a.e. in } Q \ ,$$

(A5) $$r(\mathcal{A}^{\mu+\zeta_1}) < 1 \quad ,$$

where $r(\mathcal{A}^{\mu+\zeta_1})$ is the spectral radius of the operator

$$\mathcal{A}^{\mu+\zeta_1} : L_T^\infty(\mathbf{R}) \to L_T^\infty(\mathbf{R}) \ ,$$

given by (4.9) in Section 2.4.

By Theorem 2.4.4, we conclude that for any $u \in \mathcal{U}$ there exists a unique solution to (3.1). For the sake of simplicity we shall denote this solution by p^u insteadt of $p^{\mu+u}$ as in Section 2.4.

Existence of optimal control

THEOREM 3.3.1. *Problem* (**P**) *admits at least one optimal pair.*

Proof. Consider the cost functional $\Psi : \mathcal{U} \to \mathbf{R}$, defined by

$$\Psi(u) = \int_0^T \int_0^{a_\dagger} u(a,t)p^u(a,t)da\ dt, \quad u \in \mathcal{U}$$

and let

$$d = \sup_{u \in \mathcal{U}} \Psi(u) \ .$$

Since by Theorem 2.4.4

$$0 \le \Psi(u) \le \|\zeta_2\|_{L^\infty_T(Q)} \int_0^T \int_0^{a_\dagger} p^{\zeta_1}(a,t)da\, dt < +\infty,$$

we infer that $0 \le d < +\infty$.

Now let $\{u_n\} \subset \mathcal{U},\ n \in \mathbf{N}^*$, be a sequence such that

$$d - \frac{1}{n} < \Psi(u_n) \le d\ . \tag{3.2}$$

Since by Theorem 2.4.4

$$0 \le p^{u_n}(a,t) \le p^{\zeta_1}(a,t) \quad \text{a.e. in } \mathbf{Q}\ ,$$

we conclude that there exists a subsequence, also denoted by $\{u_n\}$, such that

$$p^{u_n} \to p^* \quad \text{weakly in } L^2(Q)\ .$$

Moreover, using Mazur's Theorem we obtain a sequence $\{\tilde{p}_n\}$ satisfying

$$\tilde{p}_n = \sum_{i=n+1}^{k_n} \lambda_i^n p^{u_i}\ ,\ \lambda_i^n \ge 0\ ,\ \sum_{i=n+1}^{k_n} \lambda_i^n = 1$$

and

$$\tilde{p}_n \to p^* \quad \text{in } L^2(Q)\ . \tag{3.3}$$

Consider now the following sequence of controls

$$\tilde{u}_n(a,t) = \begin{cases} \dfrac{\sum_{i=n+1}^{k_n} \lambda_i^n p^{u_i}(a,t) u_i(a,t)}{\sum_{i=n+1}^{k_n} \lambda_i^n p^{u_i}(a,t)}, & \text{if } \sum_{i=n+1}^{k_n} \lambda_i^n p^{u_i}(a,t) \ne 0, \\ \zeta_1(a,t), & \text{if } \sum_{i=n+1}^{k_n} \lambda_i^n p^{u_i}(a,t) = 0\ . \end{cases}$$

For these controls we have

$$\tilde{u}_n \in \mathcal{U}\ ,\ \tilde{p}_n(a,t) = p^{\tilde{u}_n}(a,t) \quad \text{a.e. in } Q$$

and we can take a subsequence (also denoted by $\{\tilde{u}_n\}$) such that

$$\begin{cases} \tilde{u}_n \to u^* & \text{weakly in } L^2(Q), \\ \Psi(\tilde{u}_n) = \sum_{i=n+1}^{k_n} \lambda_i^n \Psi(u_i) \to d & \text{as } n \to +\infty\ , \end{cases} \tag{3.4}$$

(where we have used (3.2)). Thus by (3.3) and (3.4) we have shown that

$$d = \lim_{n \to +\infty} \int_0^T \int_0^{a_\dagger} \tilde{u}_n(a,t) p^{\tilde{u}_n}(a,t)da\, dt = \int_0^T \int_0^{a_\dagger} u^*(a,t) p^*(a,t)da\, dt.$$

We are left with the proof of the following equality

$$p^*(a,t) = p^{u^*}(a,t) \quad \text{a.e. in } Q\ . \tag{3.5}$$

(p^* is the solution of (3.1) coresponding to $u := u^*$). In order to prove (3.5) we recall that $p^{\tilde{u}_n}$ is a solution to (3.1) corresponding to $u := \tilde{u}_n$, in the weak sense defined in Section 2.4, that is

$$\int_0^T \int_0^{a_\dagger} [-D\varphi + (\mu(a,t) + \tilde{u}_n(a,t))\varphi - \beta(a,t)\varphi(0,t)] p^{\tilde{u}_n}(a,t)da\ dt$$

$$= \int_0^T \int_0^{a_\dagger} f(a,t)\varphi(a,t)da\ dt,$$

$\forall \varphi \in \Phi$ (with Φ defined in Section 2.4). Then by (3.3) and (3.4)

$$\int_0^T \int_0^{a_\dagger} [-D\varphi + (\mu(a,t) + u^*(a,t))\varphi - \beta(a,t)\varphi(0,t)] p^*(a,t)da\ dt$$

$$= \int_0^T \int_0^{a_\dagger} f(a,t)\varphi(a,t)da\ dt,$$

$\forall \varphi \in \Phi$, and by the uniqueness Theorem 2.4.5, (3.5) follows.

The maximum principle

Concerning the optimality conditions, the following result holds:

THEOREM 3.3.2. *Suppose that in addition $f(a,t) > 0$ a.e. in Q. If (u^*, p^*) is an optimal pair for problem* (**P**), *then there exists a unique q, solution of*

$$\begin{cases} i) \quad Dq - \mu q = u^*(1+q) - \beta(a,t)q(0,t), & (a,t) \in Q, \\ ii) \quad q(a_\dagger, t) = 0, & t \in \mathbf{R}, \\ iii) \quad q(a,t) = q(a,t+T), & (a,t) \in Q, \end{cases} \tag{3.6}$$

and in addition

$$u^*(a,t) = \begin{cases} \zeta_1(a,t), & if\ 1 + q(a,t) < 0, \\ \zeta_2(a,t), & if\ 1 + q(a,t) > 0\ . \end{cases} \tag{3.7}$$

Proof. Existence and uniqueness of q, a solution to (3.6), follows by Theorem 2.4.6.

Denote by $N_{\mathcal{U}}(u^*)$ the normal cone at $\mathcal{U}$ in u^* (in $L_T^\infty(Q)$).

Since (u^*, p^*) is an optimal pair for **(P)** we get

$$\int_0^T \int_0^{a_\dagger} u^* p^* da\, dt \geq \int_0^T \int_0^{a_\dagger} (u^* + \varepsilon v) p^{u^*+\varepsilon v} da\, dt,$$

for any $v \in L_T^\infty(Q)$, such that $u^* + \varepsilon v \in \mathcal{U}$ for any $\varepsilon > 0$ small enough, which implies

$$\int_0^T \int_0^{a_\dagger} u^* \frac{p^{u^*+\varepsilon v} - p^*}{\varepsilon} da\, dt + \int_0^T \int_0^{a_\dagger} v p^{u^*+\varepsilon v} da\, dt \leq 0. \tag{3.8}$$

Now, by Theorem 2.4.4, it immediately follows that for any v as above

$$p^{u^*+\varepsilon v} \to p^* \text{ in } L_T^\infty(Q), \text{ as } \varepsilon \to 0^+$$

and

$$\frac{p^{u^*+\varepsilon v} - p^*}{\varepsilon} \to z \text{ in } L_T^\infty(Q), \text{ as } \varepsilon \to 0^+,$$

where z is the solution of the following problem

$$\begin{cases} Dz + \mu(a,t) z = -vp^* - u^* z, & (a,t) \in Q, \\ z(0,t) = \displaystyle\int_0^{a_\dagger} \beta(a,t) z(a,t) da, & t \in \mathbf{R}, \\ z(a,t) = z(a, t+T), & (a,t) \in Q\,. \end{cases} \tag{3.9}$$

Then passing to the limit in (3.8) ($\varepsilon \to 0^+$) we conclude

$$\int_0^T \int_0^{a_\dagger} (u^* z + vp^*) da\, dt \leq 0\,. \tag{3.10}$$

Multiplying the first equation in (3.6) by z and integrating over $[0,T] \times [0, a_\dagger]$ we obtain

$$\int_0^T \int_0^{a_\dagger} (Dq - \mu q) z\, da\, dt = \int_0^T \int_0^{a_\dagger} [u^*(1+q) z - \beta(a,t) q(0,t) z] da\, dt$$

and integrating by parts and using $(3.6)_{2,3}$ and $(3.9)_{2,3}$, we get, after some calculation,

$$-\int_0^T \int_0^{a_\dagger} q(Dz + \mu z) da\, dt = \int_0^T \int_0^{a_\dagger} u^*(1+q) z\, da\, dt\,. \tag{3.11}$$

By $(3.9)_1$ and (3.11) we deduce:

$$\int_0^T \int_0^{a_\dagger} qvp^* da\, dt = \int_0^T \int_0^{a_\dagger} u^* z\, da\, dt\,. \tag{3.12}$$

Using now (3.10) and (3.12) we have that

$$\int_0^T \int_0^{a_\dagger} v(1+q)p^* da\; dt \leq 0,$$

for any $v \in L_T^\infty(Q)$, such that $u^* + \varepsilon v \in \mathcal{U}$, for any $\varepsilon > 0$ small enough. Taking into account the structure of $N_{\mathcal{U}}(u^*)$ (see Appendix 1) this implies

$$(1+q)p^* \in N_{\mathcal{U}}(u^*),$$

or equivalently (3.7) (because the strict positiveness of f implies the strict positiveness of p^* a.e. in Q).

REMARK 3.3.1. *In the case when f is strictly positive and if the following assumption hold*

$$(\mathbf{A6}) \qquad \begin{cases} \mu(a,t) > 0, \quad \text{a.e. in } Q \\ \text{and for almost any } t \in \mathbf{R}, \; \dfrac{\beta}{\mu}(\cdot,t) \\ \text{is not a strictly positive constant on any subset} \\ \text{of positive measure ,} \end{cases}$$

then $(3.6)_1$ implies that the set $\{(a,t) \in Q;\; q(a,t) = -1\}$ has the Lebesgue measure 0 (q is the solution of (3.6)). That means u^* is a bang-bang control.

About the uniqueness of the optimal control, the following result holds:

THEOREM 3.3.3. *If* **(A1)** - **(A6)** *hold and if $f(a,t) > 0$ a.e. in Q, then the optimal control is unique.*

Proof. Suppose that u_1^*, u_2^* are two optimal controls for **(P)** and denote by q_i, $i \in \{1,2\}$, the solutions of (3.6) corresponding to $u^* := u_i^*$, $i \in \{1,2\}$, respectively, and by $w = q_1 - q_2$. By (3.6) and (3.7) it follows that q_i is a solution to:

$$\begin{cases} Dq_i - (\mu + \zeta_1)q_i = \zeta_1(a,t) + (\zeta_2 - \zeta_1)(1+q_i)^+ & \\ \qquad\qquad -\beta(a,t)q_i(0,t), & (a,t) \in Q, \\ q_i(a_\dagger,t) = 0, & t \in \mathbf{R}, \\ q_i(a,t) = q_i(a,t+T), & (a,t) \in Q\;, \end{cases}$$

$i \in \{1, 2\}$ and w is a solution to

$$\begin{cases} Dw - (\mu + \zeta_1)w = (\zeta_2 - \zeta_1)[(1+q_1)^+ \quad -(1+q_2)^+] \\ \qquad -\beta(a,t)w(0,t), & (a,t) \in Q, \\ w(a_\dagger, t) = 0, & t \in \mathbf{R}, \\ w(a,t) = w(a, t+T), & (a,t) \in Q\,. \end{cases} \tag{3.13}$$

Since $|w|$ is absolutely continuous along almost every characteristic line (because w is), we deduce from (3.13) that $|w|$ is the solution to:

$$\begin{cases} Dh - (\mu + \zeta_1)h = l(a,t) - \beta(a,t)h(0,t), & (a,t) \in Q, \\ h(a_\dagger, t) = 0, & t \in \mathbf{R}, \\ h(a,t) = h(a,t+T), & (a,t) \in Q, \end{cases}$$

where

$$\begin{aligned} l(a,t) = {} & (\zeta_2 - \zeta_1)[(1+q_1)^+ - (1+q_2)^+]sign(q_1 - q_2) \\ & -\beta(a,t)w(0,t)sign(q_1 - q_2) + \beta(a,t)|w(0,t)|\,. \end{aligned}$$

It is clear that $l \in L^\infty_T(Q)$ and $l(a,t) \geq 0$ a.e. in Q. Thus by Theorem 2.4.6, it follows that $|w(a,t)| \leq 0$ a.e. in Q and consequently $q_1(a,t) = q_2(a,t)$ a.e. in Q.

Now, since

$$u_i^*(a,t) = \begin{cases} \zeta_1(a,t), & if\ 1 + q_i(a,t) < 0, \\ \zeta_2(a,t), & if\ 1 + q_i(a,t) > 0, \end{cases}$$

($i \in \{1,2\}$), $q_1(a,t) = q_2(a,t)$ a.e. in Q and $q_i \neq -1$ a.e. in Q, we may conclude that $u_1^* = u_2^*$ in $\mathcal{U}$.

REMARK 3.3.2. *Assume that the hypotheses in Theorem 3.3.3 hold. Since q (from Theorem 3.3.2) is the unique solution to*

$$\begin{cases} Dq - (\mu + \zeta_1)q = \zeta_1 + (\zeta_2 - \zeta_1)(1+q)^+ - \beta(a,t)q(0,t), & (a,t) \in Q, \\ q(a_\dagger, t) = 0, & t \in \mathbf{R}, \\ q(a,t) = q(a,t+T), & (a,t) \in Q\,, \end{cases}$$

it follows that q does not depend on the strictly positive flow f and consequently the optimal control is unique and does not depend on f.

It means that for every strictly positive flow f, the optimal control in problem **(P)** *is the same.*

Denote this optimal control by u^*.

The last remark allows us to formulate the following result for the general case $f(a,t) \geq 0$ a.e. in Q:

THEOREM 3.3.4. *If* **(A1)** - **(A6)** *hold, then the control u^* is optimal also for problem* (**P**) *corresponding to the nonnegative flow f.*

Proof. Consider $f_n \in L^\infty_T(Q)$, such that $f_n(a,t) > 0$ a.e. in Q and $f_n \to f$ in $L^\infty_T(Q)$ (as $n \to +\infty$). Denote by $\Psi_n(u)$ the cost function corresponding to $f := f_n$ in problem **(P)** and by p^u_n the solution to (3.1) corresponding to u and $f := f_n$.

Using (4) in Theorem 2.4.4, we conclude that for any $u \in \mathcal{U}$, $p^u_n \to p^u$ in $L^\infty_T(Q)$, as $n \to +\infty$. This implies

$$\Psi_n(u) \to \Psi(u) \quad \text{as } n \to +\infty \ . \tag{3.14}$$

Since for any $u \in \mathcal{U}$ we have $\Psi_n(u^*) \geq \Psi_n(u)$, and using (3.14), we conclude that u^* is an optimal control for **(P)**.

Note. By similar methods, the following problem can be studied:

$$Maximize \int_0^T \int_0^{a_\dagger} g(a,t)u(a,t)p^u(a,t)da\ dt \ ,$$

over all $u \in \mathcal{U}$, where p^u is the solution of (3.1). Here g is a weight function ($g(a,t)$ is the cost of an individual of age a, at the moment t). The integral

$$\int_0^T \int_0^{a_\dagger} g(a,t)u(a,t)p^u(a,t)da\ dt$$

represents the total cost of the harvested population (with the harvesting effort u).

The usual assumptions on g are:

$$\textbf{(A7)} \qquad \begin{cases} g \in L^\infty_T(Q), \\ g(a,t) \geq 0 \quad \text{a.e. in } Q \ . \end{cases}$$

3.4. A nonlinear optimal control problem

This section is devoted to an optimal control problem governed by a nonlinear age-dependent population dynamics.

The main goal of this section is to obtain first order necessary conditions of optimality. These conditions allow us to approximate the optimal control and the optimal cost value. We shall use as the main ingredient the variational principle of Ekeland (see Appendix 1).

Namely we deal with the following problem:

$$\textbf{(P)} \qquad Minimize \int_0^T \int_0^{a_\dagger} \left[g(p^u(a,t)) + \frac{1}{2}\rho |u(a,t)|^2 \right] da\, dt,$$

($\rho > 0$) subject to $u \in \mathcal{U} = \{v \in L^1(Q_T);\ 0 \le v(a,t) \le L \text{ a.e.}\}$, where p^u is the solution of

$$\begin{cases} Dp(a,t) + \mu_0(a)p(a,t) = -\mu(H(t))p(a,t) & (a,t) \in Q_T, \\ p(0,t) = \displaystyle\int_0^{a_\dagger} \beta(a)p(a,t)da & t \in (0,T), \\ p(a,0) = p_0(a) & a \in (0,a_\dagger), \\ H(t) = \displaystyle\int_0^{a_\dagger} u(a,t)p(a,t)da & t \in (0,T). \end{cases} \tag{4.1}$$

We suppose the following assumptions on p_0, β, μ_0 and μ hold:

$$\textbf{(as1)} \qquad \beta \in L^\infty(0,a_\dagger)\ ,\ \beta(a) \ge 0 \quad \text{a.e. in } (0,a_\dagger),$$

$$\textbf{(as2)} \qquad \begin{cases} \mu_0 \in L^1_{loc}([0,a_\dagger))\ ,\ \mu_0(a) \ge 0 & \text{a.e. in } (0,a_\dagger), \\ \displaystyle\int_0^{a_\dagger} \mu_0(a)da = +\infty, & \\ \mu \in C^2(\mathbf{R}^+),\ \mu(t) \ge 0, & \forall t \in \mathbf{R}^+, \\ \mu',\ \mu'' \text{ are bounded,} & \end{cases}$$

$$\textbf{(as3)} \qquad p_0 \in L^\infty(0,a_\dagger)\ ,\ p_0(a) \ge 0 \quad \text{a.e. in } (0,a_\dagger).$$

By *a solution* to (4.1) we mean a function $p^u \in L^\infty(Q_T)$, absolutely continuous along almost every characteristic line, which satisfies

$$\begin{cases} Dp^u(a,t) + \mu_0(a)p^u(a,t) = -\mu(H^u(t))p^u(a,t) & \text{a.e. in } Q_T, \\ \lim_{\varepsilon\to 0+} p^u(\varepsilon, t+\varepsilon) = \int_0^{a_\dagger} \beta(a)p^u(a,t)da & \text{a.e. } t \in (0,T), \\ \lim_{\varepsilon\to 0+} p^u(a+\varepsilon,\varepsilon) = p_0(a) & \text{a.e. } a \in (0,a_\dagger), \end{cases}$$

where $H^u(t) = \int_0^{a_\dagger} u(a,t)p^u(a,t)da$ a.e. in $(0,T)$.

In the same manner as in Section 2.2, we can state the following result:

LEMMA 3.4.1. *For any $u \in \mathcal{U}$ there exists a unique solution p^u to (4.1). The solution is nonnegative.*

Moreover, there exists a function $C(M,T)$ such that if $\|p_0\|_{L^\infty(0,a_\dagger)} \le M$, then

$$\|p^u(t) - p^v(t)\|_{L^1(0,a_\dagger)} \le C(M,T)\|p_0\|_{L^\infty(0,a_\dagger)} \int_0^t \|u(s) - v(s)\|_{L^1(0,a_\dagger)}ds,$$

a.e. $t \in (0,T)$.

The following assumptions on g will be in effect throughout this section,

(as4)
$$\begin{cases} g \in C^2(\mathbf{R}^+),\ g(x) \ge 0,\ \forall x \in \mathbf{R}^+, \\ g',\ g'' \text{ are bounded.} \end{cases}$$

OPTIMALITY CONDITIONS

Consider the mapping $\mathcal{P} : L^1(Q_T) \to L^\infty(Q_T)$, defined by

$$\mathcal{P}(f) = \begin{cases} 0, & \text{if } f(a,t) \le 0, \\ f(a,t), & \text{if } 0 < f(a,t) < L, \\ L, & \text{if } f(a,t) \ge L. \end{cases}$$

THEOREM 3.4.2. *Let u^* be an optimal control for* (**P**). *If q^{u^*} is the solution to*

$$\begin{cases} Dq(a,t) - \mu_0(a)q(a,t) - \mu(H^u(t))q(a,t) + \beta(a)q(0,t) \\ -u(a,t)\mu'(H^u(t)) \int_0^{a_\dagger} p^u(a,t)q(a,t)da = g'(p^u(a,t)), & (a,t) \in Q_T, \\ q(a_\dagger,t) = 0 & t \in (0,T), \\ q(a,T) = 0 & a \in (0,a_\dagger), \\ H^u(t) = \int_0^{a_\dagger} u(a,t)p^u(a,t)da & t \in (0,T), \end{cases} \tag{4.2}$$

corresponding to $u := u^*$, then

$$u^*(a,t) = \mathcal{P}\left[-\rho^{-1}p^{u^*}(a,t)\mu'(H^{u^*}(t)) \int_0^{a_\dagger} p^{u^*}(a,t)q^{u^*}(a,t)da\right]. \tag{4.3}$$

Proof. Using a fixed point result follows the existence of a unique solution to (4.2) (as in Section 2.2).

Denote by $\mathcal{V}$ the set of all $v \in L^\infty(Q_T)$ such that

$$\begin{cases} v(a,t) \geq 0, & \text{if } u^*(a,t) = 0, \\ v(a,t) \leq 0, & \text{if } u^*(a,t) = L. \end{cases}$$

For any $v \in \mathcal{V}$ and for any $\lambda > 0$ small enough we have that

$$\int_0^T \int_0^{a_\dagger} \left[g(p^{u^*}(a,t)) + \frac{1}{2}\rho|u^*(a,t)|^2\right] da\ dt$$

$$\leq \int_0^T \int_0^{a_\dagger} \left[g(p^{u^*+\lambda v}(a,t)) + \frac{1}{2}\rho|u^*(a,t) + \lambda v(a,t)|^2\right] da\ dt.$$

This implies that

$$\int_0^T \int_0^{a_\dagger} \left[g'(p^{u^*}(a,t))z(a,t) + \rho u^*(a,t)v(a,t)\right] da\ dt \geq 0, \tag{4.4}$$

for any $v \in \mathcal{V}$, where z is the solution of the system

$$Dz(a,t) + \mu_0(a)z(a,t) + \mu(H^{u^*}(t))z(a,t)$$

$$+\mu'(H^{u^*}(t))p^{u^*}(a,t)\int_0^{a_\dagger}[v(a,t)p^{u^*}(a,t)+u^*(a,t)z(a,t)]da=0$$
$$(a,t)\in Q_T,$$

$$z(0,t)=\int_0^{a_\dagger}\beta(a)z(a,t)da \qquad t\in(0,T),$$

$$z(a,0)=0 \qquad a\in(0,a_\dagger).$$

z satisfies

$$z(a,t)=\lim_{\lambda\to 0^+}\frac{p^{u^*+\lambda v}(a,t)-p^{u^*}(a,t)}{\lambda} \qquad \text{a.e. } (a,t)\in Q_T.$$

Note that the last system and (4.2) assure that

$$\int_{Q_T} Dq^{u^*}z\,da\,dt=-\int_{Q_T} q^{u^*}Dz\,da\,dt-\int_0^T q^{u^*}(0,t)z(0,t)dt$$

and

$$\int_{Q_T} u^*(a,t)z(a,t)\mu'(H^{u^*}(t))\int_0^{a_\dagger}p^{u^*}(\alpha,t)q^{u^*}(\alpha,t)d\alpha\,da\,dt$$
$$=\int_{Q_T}\mu'(H^{u^*}(t))p^{u^*}(\alpha,t)q^{u^*}(\alpha,t)\int_0^{a_\dagger}u^*(a,t)z(a,t)da\,d\alpha\,dt.$$

Using these equalities and (4.2) we obtain that

$$\int_{Q_T} g'(p^{u^*}(a,t))z(a,t)da\,dt$$
$$=\int_{Q_T} v(a,t)p^{u^*}(a,t)\mu'(H^{u^*}(t))\int_0^{a_\dagger}p^{u^*}(\alpha,t)q^{u^*}(\alpha,t)d\alpha\,da\,dt.$$

Using (4.4) we see that

$$\int_{Q_T}\left[\rho u^*(a,t)+p^{u^*}(a,t)\mu'(H^{u^*}(t))\int_0^{a_\dagger}p^{u^*}(\alpha,t)q^{u^*}(\alpha,t)d\alpha\right]da\,dt\geq 0,$$

for any $v\in\mathcal{V}$ and in conclusion (4.3) holds.

EXISTENCE OF AN OPTIMAL CONTROL

Let $\Psi: L^1(Q_T)\to(-\infty,+\infty]$, be defined by

$$\Psi(u)=\begin{cases}\displaystyle\int_0^T\int_0^{a_\dagger}\left[g(p^u(a,t))+\frac{1}{2}\rho|u(a,t)|^2\right]da\,dt, & u\in\mathcal{U}\\[2ex] +\infty, & \text{otherwise.}\end{cases}$$

LEMMA 3.4.3. *The function* Ψ *is lower semicontinuous.*

Proof. Let $u_n \to u$ in $L^1(Q_T)$. Without any loss of generality we assume that $u_n \in \mathcal{U}$. Then $\{p^{u_n}\}$ is bounded in $L^\infty(Q_T)$ (by 0 and p^0; p^0 is the solution of (4.1) corresponding to $\mu := 0$; see Theorem 2.1.2).

Lemma 3.4.1 implies that

$$p^{u_n} \to p^u \quad \text{in } L^\infty(0,T;L^1(0,a_\dagger)).$$

Hence, on a subsequence, also denoted by $\{(u_n, p^{u_n})\}$, we have that

$$g(p^{u_n}(a,t)) + \frac{1}{2}\rho|u_n(a,t)|^2 \to g(p^u(a,t)) + \frac{1}{2}\rho|u(a,t)|^2, \quad \text{a.e. in } Q_T$$

(by (**as4**)). Using now Fatou's lemma we conclude that on a subsequence, also denoted by $\{u_n\}$ we have

$$\liminf_{n\to+\infty} \int_0^T \int_0^{a_\dagger} \left[g(p^{u_n}(a,t)) + \frac{1}{2}\rho|u_n(a,t)|^2\right] da\ dt$$

$$\geq \int_0^T \int_0^{a_\dagger} \left[g(p^u(a,t)) + \frac{1}{2}\rho|u(a,t)|^2\right] da\ dt$$

and consequently

$$\liminf_{n\to+\infty} \Psi(u_n) \geq \Psi(u).$$

This allows us to conclude that Ψ is lower semicontinuous.

According to Ekeland's variational principle (see Appendix 1) we may conclude that for every $\varepsilon > 0$, there exists $u_\varepsilon \in \mathcal{U}$ such that

$$\Psi(u_\varepsilon) \leq \inf \Psi + \varepsilon,$$

$$\Psi(u_\varepsilon) \leq \inf \{\Psi(u) + \varepsilon^{\frac{1}{2}}\|u - u_\varepsilon\|_{L^1(Q_T)};\ u \in L^1(Q_T)\}. \tag{4.5}$$

Since u_ε is a minimum point for Ψ_ε (where $\Psi_\varepsilon(u) = \Psi(u) + \varepsilon^{\frac{1}{2}}\|u - u_\varepsilon\|_{L^1(Q_T)}$) we may infer (as in the previous sections) that

$$\int_{Q_T} [\rho u_\varepsilon(a,t) - J(u_\varepsilon,t)p^{u_\varepsilon}(a,t)]\, v(a,t) da\ dt$$

$$+\varepsilon^{\frac{1}{2}} \int_{Q_T} |v(a,t)| da\ dt \geq 0,$$

for any $v \in \mathcal{V}_\varepsilon = \{h \in L^\infty(Q_T);\ h(a,t) \geq 0 \text{ if } u_\varepsilon(a,t) = 0,\ h(a,t) \leq 0 \text{ if } u_\varepsilon(a,t) = L\}$. Here

$$J(u,t) = -\mu'\left(\int_0^{a_\dagger} u(a,t)p^u(a,t)da\right)\int_0^{a_\dagger} q^u(a,t)p^u(a,t)da,$$

where q^u is the solution to (4.2).

We postpone for the time being the proof of the following auxiliary result:

LEMMA 3.4.4. *If $z \in L^1(Q_T)$ satisfies*

$$\int_{Q_T} [z(a,t)v(a,t) + \gamma|v(a,t)|]\, da\, dt \geq 0,$$

for any $v \in \mathcal{V}^u = \{h \in L^\infty(Q_T); h(a,t) \geq 0 \text{ if } u(a,t) = 0, h(a,t) \leq 0 \text{ if } u(a,t) = L\}$, then there exists $\zeta \in L^\infty(Q_T)$ such that $\|\zeta\|_{L^\infty(Q_T)} \leq 1$ and $u + \gamma\zeta \in N_{\mathcal{U}}(u)$.

Using now Lemma 3.4.4 we have that there exists $\zeta_\varepsilon \in L^\infty(Q_T)$, $\|\zeta_\varepsilon\|_{L^\infty(Q_T)} \leq 1$, such that

$$\rho u_\varepsilon - J(u_\varepsilon,t)p^{u_\varepsilon} - \varepsilon^{\frac{1}{2}}\zeta_\varepsilon \in N_{\mathcal{U}}(u_\varepsilon).$$

We may conclude that

$$u_\varepsilon(a,t) = \mathcal{P}\left[\rho^{-1}\left(J(u_\varepsilon,t)p^{u_\varepsilon}(a,t) + \varepsilon^{\frac{1}{2}}\zeta_\varepsilon(a,t)\right)\right], \text{ a.e. in } Q_T. \tag{4.6}$$

In the same manner as in Lemma 3.4.1 we can state

LEMMA 3.4.5. *There exists a function $C_1(M,T)$ such that if $\|p_0\|_{L^\infty(0,a_\dagger)} \leq M$, then*

$$\|q^u(t) - q^v(t)\|_{L^1(0,a_\dagger)} + |q^u(0,t) - q^v(0,t)|$$
$$\leq C_1(M,T)\|p_0\|_{L^\infty(0,a_\dagger)}\|u(s) - v(s)\|_{L^\infty(0,a_\dagger)}$$

a.e. $t \in (0,T)$.

The existence of an optimal control is clarified by

THEOREM 3.4.6. *If*

$$TC_1(M,T)\|p_0\|_{L^\infty(0,a_\dagger)}\rho^{-1} < 1,$$

then problem (**P**) *has a unique optimal control* $u^* \in \mathcal{U}$.

Proof. We first prove the uniqueness. Define the mapping $\mathcal{F} : \mathcal{U} \subset L^\infty(Q_T) \to \mathcal{U}$,

$$(\mathcal{F}u)(a,t) = \mathcal{P}(\rho^{-1}J(u,t)p^u(a,t)), \quad \text{a.e. in } Q_T.$$

We have that

$$\|\mathcal{F}u - \mathcal{F}v\|_{L^\infty(Q_T)} \leq \|p_0\|_{L^\infty(0,a_\dagger)}\rho^{-1}C_1(M,T)T\|u-v\|_{L^\infty(Q_T)}$$

and so, the hypothesis of our theorem implies via Lemma 3.4.5 that $\mathcal{F}$ is a contraction and has a unique fixed point $\tilde{u}$. By (4.3) it follows that this fixed point is unique.

In order to prove the existence of an optimal control we shall prove that $\tilde{u}$ minimizes Ψ. We have (by (4.5) and (4.6))

$$|(\mathcal{F}u_\varepsilon)(a,t) - \mathcal{P}(\rho^{-1}(J(u_\varepsilon,t)p^{u_\varepsilon}(a,t) + \varepsilon^{\frac{1}{2}}\zeta_\varepsilon(a,t)))|$$
$$\leq \varepsilon^{\frac{1}{2}}|\zeta_\varepsilon(a,t)| \leq \varepsilon^{\frac{1}{2}}\rho^{-1},$$

and consequently

$$\|\tilde{u} - u_\varepsilon\|_{L^\infty(Q_T)} \leq \|\mathcal{F}\tilde{u} - \mathcal{F}u_\varepsilon\|_{L^\infty(Q_T)} + \varepsilon^{\frac{1}{2}}\rho^{-1}.$$

This implies that $u_\varepsilon \to \tilde{u}$ in $L^\infty(Q_T)$ and that

$$\Psi(\tilde{u}) = \inf_{u\in\mathcal{U}} \Psi(u).$$

We shall prove now Lemma 3.4.4. Let

$$A_0 = \{(a,t) \in Q_T;\ 0 < u(a,t) < L\},$$

$$A_1 = \{(a,t) \in Q_T;\ u(a,t) = 0\},$$

$$A_2 = \{(a,t) \in Q_T;\ u(a,t) = L\}.$$

The hypothesis implies that

$$\int_{Q_T} [z(a,t) + \gamma]v(a,t)da\ dt \geq 0,$$

for any $v \in L^\infty(Q_T)$ such that

$$v(a,t) \geq 0 \qquad \text{on } A_1,$$

$$v(a,t) = 0 \qquad \text{on } Q_T \setminus A_1.$$

This implies that

$$z(a,t) + \gamma \geq 0 \qquad \text{a.e. on } A_1.$$

In the same manner we can obtain that

$$z(a,t) - \gamma \leq 0, \qquad \text{a.e. on } A_2$$

and

$$|z(a,t)| \leq \gamma, \qquad \text{a.e. on } A_0.$$

The last three relations imply the conclusion of the lemma.

Consider now the following optimal control problem:

$$\textbf{(P1)} \qquad \textit{Minimize} \int_0^T \int_0^{a_\dagger} [g(p^u(a,t) - \overline{p}(a,t)) + h(u(a,t))]\, da\, dt,$$

subject to $u \in L^2(Q_T)$, where p^u is the solution of

$$\begin{cases} Dp(a,t) + \mu_0(a)p(a,t) = u(a,t), & (a,t) \in Q_T, \\ p(0,t) = \displaystyle\int_0^{a_\dagger} \beta(a)p(a,t)da, & t \in (0,T), \\ p(a,0) = p_0(a), & a \in (0,a_\dagger). \end{cases}$$

Here $g,\ h : \mathbf{R} \to \mathbf{R}^+$ are convex functions and $\overline{p} \in L^\infty(Q_T)$.

We propose to the reader to prove that this problem admits at least one optimal pair (u^*, p^*). Moreover, if q is the solution to

$$\begin{cases} Dq - \mu_0(a)q + \beta(a)q(0,t) \in \partial g(p^*(a,t) - \overline{p}(a,t)), & (a,t) \in Q_T, \\ q(a_\dagger, t) = 0 & t \in (0,T), \\ q(a,T) = 0 & a \in (0,a_\dagger), \end{cases}$$

then

$$q(a,t) \in \partial h(u^*(a,t)), \qquad \text{a.e. } (a,t) \in Q_T.$$

This result follows by the scheme presented at the beginning of this chapter (for regularization of a convex function see Appendix 1).

Anyway, an important particular situation of **(P1)** is that when $g(x) = |x|^2$ and h is the indicator function of $[0,L]$, or $h(u) = |u|^2$.

*

* *

The optimal control of age-dependent population dynamics has been intensively studied in the last two decades. M. Brokate [24], [25] published two of the first important papers on this topic. Great attention has been given to the optimal harvesting problem. In this regard we refer also to the fundamental papers of M.E. Gurtin and L.F. Murphy [46], [47] and of L.F. Murphy and S.J. Smith [75]. The main attention is paid to the necessary optimality conditions (the maximum principle) and to approximation of the optimal control. The approximation of the optimal harvesting effort has been approached by S. Aniţa [6]. A time-periodic harvesting problem was recently investigated by S. Aniţa, M. Iannelli, M.-Y. Kim and E.-J. Park [11]. As concerns some other optimal control problems related to age-dependent population dynamics, we refer to the papers of S. Aniţa [6], V. Barbu and M. Iannelli [19] and of W.L. Chan and B.Z. Guo [28], [29].

The approach in Section 3.1 is inspired by the paper of M. Brokate [24].

The results in the next section were published by the author in [10].

The study of periodic age-dependent population dynamics performed in Section 3.3 is based on [11].

The results in the last section concern some other optimal control problems. They were obtained by V. Barbu and M. Iannelli [19], who in fact have investigated in [19] a more general problem.

3.5. Exercises

Consider the optimal control problem:

$$\textbf{(E1)} \qquad Maximize \int_0^T \int_0^1 u(t)p^u(a,t)da\, dt,$$

subject to $u \in L^2(0,T)$, $0 \le u(t) \le L$ a.e., where p^u is the solution to

$$\begin{cases} Dp + p = -u(t)p, & (a,t) \in (0,1) \times (0,T), \\ p(0,t) = \int_0^1 p(a,t)da, & t \in (0,T), \\ p(a,0) = 1, & a \in (0,1) \end{cases}$$

$(L > 0)$.

1. Prove that problem **(E1)** admits at least one optimal pair.

Hint. The cost function I_1 is a lower semicontinuous convex proper function. In addition it satisfies

$$\lim_{\|u\|_{L^2(0,T)} \to +\infty} I_1(u) = +\infty.$$

By Theorem A1.1.3 it follows that there exists at least one minimum point for I_1.

2. Prove that if u^* is an optimal control for **(E1)**, then $u^* = L$ a.e. $t \in (0,T)$.

Hint. Denote by $\tilde{p}$ the solution of

$$\begin{aligned} &Dp + p = 0, && (a,t) \in (0,1) \times (0,T), \\ &p(0,t) = \int_0^1 p(a,t)da, && t \in (0,T), \\ &p(a,0) = 1, && a \in (0,1). \end{aligned}$$

It is obvious that $\tilde{p} \equiv 1$. Problem **(E1)** is equivalent to the following one:

$$Minimize \int_0^T u(t)\tilde{P}(t)y^u(t)dt,$$

where $\tilde{P}(t) = \displaystyle\int_0^1 \tilde{p}(a,t)da = 1$ and y^u is the solution to

$$\begin{aligned} &y'(t) = -u(t)y(t), && t \in (0,T), \\ &y(0) = 1. \end{aligned}$$

The optimality conditions for this last problem allow us to get the conclusion.

Consider the optimal control problem

$$\textbf{(E2)} \qquad Minimize \int_0^T \int_0^{a_\dagger} |p^u(a,t) - p^0(a)|^2 da\, dt,$$

subject to $u \in L^2((0,a_\dagger) \times (0,T))$, $0 \leq u(a,t) \leq L$ $(L > 0)$ a.e., where p^u is the solution to

$$\begin{aligned} &Dp + \mu(a,t)p = u(a,t), && (a,t) \in (0,a_\dagger) \times (0,T), \\ &p(0,t) = \int_0^{a_\dagger} \beta(a,t)p(a,t)da, && t \in (0,T), \\ &p(a,0) = p_0(a), && a \in (0,a_\dagger). \end{aligned}$$

Assume that the hypotheses in Section 2.1 are satisfied and that

$$p^0 \in L^\infty((0,a_\dagger) \times (0,T)).$$

3. Show that problem **(E2)** admits at least one optimal pair.

Hint. The cost function I_2 is a lower semicontinuous convex proper function. In addition it satisfies

$$\lim_{\|u\|_{L^2((0,a_\dagger)\times(0,T))}\to+\infty} I_2(u) = +\infty.$$

By Theorem A1.1.3 it follows that there exists at least one minimum point for I_2.

4. Show that if (u^*, p^*) is an optimal pair for **(E2)**, then

$$u^*(a,t) = \begin{cases} 0, & \text{if } q(a,t) < 0, \\ L, & \text{if } q(a,t) > 0, \end{cases}$$

where q is the solution to

$$\begin{cases} Dq - \mu q = -\beta(a,t)q(0,t) + 2(p^* - p^0), & (a,t) \in Q_T, \\ q(a_\dagger, t) = 0, & t \in (0,T), \\ q(a,T) = 0, & a \in (0,a_\dagger). \end{cases}$$

Consider the optimal harvesting problem

$$\textbf{(E3)} \qquad Maximize \int_0^{2\pi}\int_0^1 u(a)p^u(a,t)da\, dt,$$

subject to $u \in L^\infty(0,1)$, $0 \le u(a) \le L$ a.e., where p^u is the solution to

$$\begin{aligned} & Dp(a,t) + p(a,t) = 1 - \cos t - u(a)p(a,t), && (a,t) \in (0,1) \times \mathbf{R}, \\ & p(0,t) = \int_0^1 p(a,t)da, && t \in \mathbf{R}, \\ & p(a,t) = p(a,t+2\pi), && (a,t) \in (0,1) \times \mathbf{R} \end{aligned}$$

$(L > 0)$.

5. Prove that **(E3)** admits at least one optimal control.

Hint. Denote by

$$P^u(a) = \int_0^{2\pi} p^u(a,s)ds = \int_t^{t+2\pi} p^u(a,s)ds,$$

$a \in [0,1]$, $t \in \mathbf{R}$. Remark that

$$\int_0^{2\pi} \int_0^1 u(a)p^u(a,t)da\ dt = \int_0^1 u(a)P^u(a)da.$$

Problem **(E3)** is equivalent to the following one:

$$Maximize \int_0^1 u(a)P^u(a)\ da,$$

subject to $u \in L^\infty(0,1)$, $0 \le u(a) \le L$ a.e., where P^u is the solution to

$$P'(a) + P(a) = 2\pi - u(a)P(a), \quad a \in (0,1),$$

$$P(0) = \int_0^1 P(a)da.$$

The existence of an optimal control for this last problem follows as in Section 3.1.

6. Show that if (u^*, p^*) is an optimal pair for **(E3)** then

$$u^* = \begin{cases} 0, & \text{if } 1 + q(a) < 0, \\ L, & \text{if } 1 + q(a) > 0, \end{cases} \tag{5.1}$$

where q is the solution to

$$\begin{aligned} &q'(a) - q(a) = u^*(a)(1 + q(a)) - q(0), \quad a \in (0,1), \\ &q(1) = 0. \end{aligned} \tag{5.2}$$

Hint. In the same manner as in Section 3.1 it can be proved that u^* satisfies (5.1), where q is the solution to (5.2). Using now (5.1) and (5.2) the conclusion follows.

Consider the optimal harvesting problem:

$$\textbf{(E4)} \qquad Maximize \int_0^T \int_0^1 up^u(a,t)da\ dt,$$

subject to $u \in [0, L]$ (here the control is constant), where p^u is the solution to

$$Dp(a,t) + \frac{1}{1-a}p(a,t) = -up(a,t), \qquad (a,t) \in (0,1) \times (0,T),$$

$$p(0,t) = 2\int_0^1 p(a,t)da, \qquad t \in (0,T),$$

$$p(a,0) = 1 - a, \qquad a \in (0,1)$$

($L > 0$).

7. Find the optimal effort and the optimal harvest.

Hint. The solution p^u is given by

$$p^u(a,t) = e^{-ut}(1-a), \quad (a,t) \in (0,1) \times (0,T)$$

and the cost value is

$$\int_0^T \int_0^1 up^u(a,t)da\, dt = \frac{1}{2}\int_0^T ue^{-ut}ds$$
$$= \frac{1}{2}(1 - e^{-uT}).$$

CHAPTER 4

ANALYSIS OF POPULATION DYNAMICS WITH DIFFUSION

This chapter is devoted to the study of age-dependent population dynamics with diffusion. The models to be studied describe a biological population which is free to move in a habitat $\Omega \subset \mathbf{R}^n$ ($n \in \{1, 2, 3\}$). The habitat is a nonempty bounded domain with a smooth enough boundary $\partial\Omega$. For the reader's convenience we do not take the most general hypotheses. However we choose a sufficiently large framework in order to cover the most important aspects related to models with diffusion. The study concerns models with homogeneous Neumann boundary conditions. By similar methods we can treat models with homogeneous Dirichlet boundary conditions.

Throughout this chapter $a_\dagger$, T, $k \in (0, +\infty)$, $Q_T = (0, a_\dagger) \times (0, T) \times \Omega$, $\Sigma_T = (0, a_\dagger) \times (0, T) \times \partial\Omega$ and Dp is the directional derivative

$$Dp(a, t, x) = \lim_{\varepsilon \to 0} \frac{p(a + \varepsilon, t + \varepsilon, x) - p(a, t, x)}{\varepsilon}.$$

4.1. Basic properties of the solution. The linear model

This section concerns the most important properties of the solution of linear age-dependent population dynamics with diffusion. Existence, uniqueness and positivity of the solution will be proved. Some comparison results for the linear model will also be established.

Consider the linear model which describes the evolution of an age structured population with diffusion

$$\begin{cases} Dp(a,t,x) + \mu(a,t,x)p(a,t,x) - k\Delta p(a,t,x) = f(a,t,x), & \text{in } Q_T, \\ \dfrac{\partial p}{\partial \nu}(a,t,x) = 0, & \text{on } \Sigma_T, \\ p(0,t,x) = \displaystyle\int_0^{a_\dagger} \beta(a,t,x)p(a,t,x)da, & \text{in } (0,T) \times \Omega, \\ p(a,0,x) = p_0(a,x), & \text{in } (0,a_\dagger) \times \Omega. \end{cases} \tag{1.1}$$

Assume that β, μ, f and p_0 satisfy the following hypotheses:

(A1) $\beta \in L^{\infty}(Q_T), \quad \beta(a,t,x) \geq 0$ a.e. in Q_T;

(A2)
$$\begin{cases} \mu \in L^{\infty}_{loc}([0,a_\dagger) \times [0,T] \times \overline{\Omega}), & \\ \mu(a,t,x) \geq \mu_0(a,t) \geq 0, & \text{a.e. in } Q_T, \\ \text{where} & \\ \mu_0 \in L^{\infty}_{loc}([0,a_\dagger) \times [0,T]) & \\ \text{and} & \\ \int_0^{a_\dagger} \mu_0(a, t - a_\dagger + a)da, & \text{a.e. } t \in (0,T) \\ (\text{here } \mu_0 \text{ is extended by zero on} & (0,a_\dagger) \times (-\infty, 0)); \end{cases}$$

(A3)
$$\begin{cases} p_0 \in L^2((0,a_\dagger) \times \Omega), \quad p_0(a,x) \geq 0 & \text{a.e. in } (0,a_\dagger) \times \Omega \\ f \in L^2(Q_T), \qquad f(a,t,x) \geq 0 & \text{a.e. in } Q_T. \end{cases}$$

By *a solution* to system (1.1) we mean a function $p \in L^2(Q_T)$, which belongs to $C(\overline{S}; L^2(\Omega)) \cap AC(S; L^2(\Omega)) \cap L^2(S; H^1(\Omega)) \cap L^2_{loc}(S; H^2(\Omega))$, for almost any characteristic line S of equation

$$a - t = a_0 - t_0, \quad (a,t) \in (0,a_\dagger) \times (0,T),$$

$(a_0, t_0) \in \{0\} \times (0,T) \cup (0,a_\dagger) \times \{0\}$ and satisfies

$$\begin{cases} Dp(a,t,x) + \mu(a,t,x)p(a,t,x) - k\Delta p(a,t,x) & = f(a,t,x), \\ & \text{a.e. in } Q_T, \\ \dfrac{\partial p}{\partial \nu}(a,t,x) = 0, & \text{a.e. in } \Sigma_T, \\ \lim\limits_{\varepsilon \to 0^+} p(\varepsilon, t+\varepsilon, \cdot) = \displaystyle\int_0^{a_\dagger} \beta(a,t,\cdot)p(a,t,\cdot)da, & \text{in } L^2(\Omega), \\ & \text{a.e. } t \in (0,T), \\ \lim\limits_{\varepsilon \to 0^+} p(a+\varepsilon, \varepsilon, \cdot) = p_0(a, \cdot), & \text{in } L^2(\Omega), \\ & \text{a.e. } a \in (0,a_\dagger). \end{cases} \tag{1.2}$$

For the characteristic line S we can write

$$S = \{(a,t) \in (0,a_\dagger) \times (0,T);\ a - t = a_0 - t_0\} = \{(a_0 + s, t_0 + s);\ s \in (0,\alpha)\}.$$

Here $(a_0 + \alpha, t_0 + \alpha) \in \{a_\dagger\} \times (0,T) \cup (0,a_\dagger) \times \{T\}$.

We have denoted by

$$C(\overline{S}; L^2(\Omega)) = \{h : \overline{S} \to L^2(\Omega);\ h \text{ continous}\},$$

$$AC(S;L^2(\Omega)) = \{h : S \to L^2(\Omega);\ h(a_0+\cdot, t_0+\cdot) : (0,\alpha) \to L^2(\Omega)$$
$$\text{is absolutely continuous on any compact subinterval}\}.$$

Since the solution to (1.1) satisfies $p \in C(\overline{S};L^2(\Omega))$, then $(1.2)_{3,4}$ are meaningful.

For the fundamentals of Sobolev spaces we refer to Adams [1].

First we shall study the case when μ satisfies

(A2)' $\qquad \mu \in L^\infty(Q_T) \quad , \quad \mu(a,t,x) \geq 0 \ a.e. \ in \ Q_T$

(in place of **(A2)**).

The following lemma is basic:

LEMMA 4.1.1. *Under the hypotheses* **(A1)**, **(A2)'**, **(A3)**, *there exists a unique solution p of (1.1).*

If

$$\begin{cases} \beta_1,\beta_2 & \text{satisfy } \mathbf{(A1)}, \\ \mu_1,\mu_2 & \text{satisfy } \mathbf{(A2)'}, \\ f_1, f_2 & \text{and } p_{01}, p_{02} \text{ satisfy } \mathbf{(A3)}, \end{cases}$$

and if

$$\beta_1 \leq \beta_2,\ \mu_1 \geq \mu_2,\ f_1 \leq f_2,\ p_{01} \leq p_{02},$$

then

$$0 \leq p_1\,(a,t,x) \leq p_2\,(a,t,x) \ \text{ a.e. in } Q_T$$

where p_i is the solution of (1.1) corresponding to $:= (\beta_i, \mu_i, f_i, p_{0i})$, $i \in \{1,2\}$.

Proof. Consider $b \in L^2((0,T)\times\Omega)$, $q \in L^2(Q_T)$ arbitrary but fixed, and let us study the following system

$$\begin{cases} Dp + \mu q - k\Delta p = f, & (a,t,x) \in Q_T, \\ \dfrac{\partial p}{\partial \nu}(a,t,x) = 0, & (a,t,x) \in \Sigma_T, \\ p(0,t,x) = b(t,x), & (t,x) \in (0,T)\times\Omega, \\ p(a,0,x) = p_0(a,x), & (a,x) \in (0,a_\dagger)\times\Omega. \end{cases} \tag{1.3}$$

We can view (1.3) as a collection of linear parabolic systems (on the characteristic lines S).

For any function $h : Q_T \to \mathbf{R}$ and for any characteristic line

$$S = \{(a_0 + s, t_0 + s)\,;\, s \in (0, \alpha)\},$$

we define

$$\widetilde{h}_S(s, x) = h(a_0 + s, t_0 + s, x), \qquad (s, x) \in (0, \alpha) \times \Omega.$$

So, let $\widetilde{p}_S \in C([0, \alpha]; L^2(\Omega)) \cap AC((0, \alpha); L^2(\Omega)) \cap L^2(0, \alpha; H^1(\Omega)) \cap L^2_{loc}((0, \alpha); H^2(\Omega))$ be the solution of

$$\begin{cases} \dfrac{\partial \widetilde{p}}{\partial s} - k\Delta \widetilde{p} = \widetilde{f}_S - \widetilde{\mu}_S \widetilde{q}_S, & (s, x) \in (0, \alpha) \times \Omega, \\ \dfrac{\partial \widetilde{p}}{\partial \nu}(s, x) = 0, & (s, x) \in (0, \alpha) \times \partial\Omega, \\ \widetilde{p}\,(0, x) = \begin{cases} b\,(t_0, x)\,, & \text{if } \ a_0 = 0 \\ p_0\,(a_0, x)\,, & \text{if } \ t_0 = 0 \end{cases}, & x \in \Omega. \end{cases} \tag{1.4}$$

By Lemma A2.7 it follows that (1.4) admits a unique solution. Multiplying $(1.4)_1$ by $\widetilde{p}_S$ and integrating over $(0, s) \times \Omega$ we obtain that

$$\|\widetilde{p}_S(s)\|^2_{L^2(\Omega)} \le \|\widetilde{p}_S(0)\|^2_{L^2(\Omega)} + \|\widetilde{f}_S - \widetilde{\mu}_S \widetilde{q}_S\|^2_{L^2((0,\alpha)\times\Omega)} + \int_0^s \|\widetilde{p}_S(\tau)\|^2_{L^2(\Omega)} d\tau,$$

and by Bellman's lemma we get

$$\|\widetilde{p}_S(s)\|^2_{L^2(\Omega)} \le (\|\widetilde{p}_S(0)\|^2_{L^2(\Omega)} + \|\widetilde{f}_S - \widetilde{\mu}_S \widetilde{q}_S\|^2_{L^2((0,\alpha)\times\Omega)}) e^{\alpha}, \tag{1.5}$$

$\forall s \in [0, \alpha]$. If we denote by

$$p_{b,q}(a_0 + s, t_0 + s, x) = \widetilde{p}_S(s, x), \qquad (s, x) \in (0, \alpha) \times \Omega,$$

for any characteristic line S, then we obtain by Lemma A2.7 and (1.5) that $p_{b,q} \in C(\overline{S}; L^2(\Omega)) \cap AC(S; L^2(\Omega)) \cap L^2(S; H^1(\Omega)) \cap L^2_{loc}(S; H^2(\Omega))$, for almost any characteristic line S, and satisfies

$$\begin{cases} Dp_{b,q} + \mu q - k\Delta p_{b,q} = f, & \text{a.e. in } Q_T, \\ \dfrac{\partial p_{b,q}}{\partial \nu}(a, t, x) = 0, & \text{a.e. in } \Sigma_T, \\ p_{b,q}(0, t, x) = b(t, x), & \text{a.e. in } (0, T) \times \Omega, \\ p_{b,q}(a, 0, x) = p_0(a, x), & \text{a.e. in } (0, a_\dagger) \times \Omega, \end{cases} \tag{1.4$'$}$$

where

$$p_{b,q}(0,t,\cdot) = \lim_{\varepsilon \to 0+} p_{b,q}(\varepsilon, t+\varepsilon, \cdot)$$

in $L^2(\Omega)$, a.e. $t \in (0,T)$,

$$p_{b,q}(a,0,\cdot) = \lim_{\varepsilon \to 0+} p_{b,q}(a+\varepsilon, \varepsilon, \cdot)$$

in $L^2(\Omega)$, a.e. $a \in (0,a_\dagger)$.

In order to prove that $p_{b,q}$ is the solution to (1.4)′ it suffices to show that $p_{b,q} \in L^2(Q_T)$. This yields

$$p_{b,q}(a,t) = \sum_{j=0}^{\infty} p^j_{b,q}(a,t)\varphi_j$$

in $L^2(\Omega)$, a.e. $(a,t) \in (0,a_\dagger) \times (0,T)$ (by Lemma A2.7), where

$$\begin{cases} Dp^j_{b,q} + \lambda_j k p^j_{b,q} = v^j(a,t), & (a,t) \in (0,a_\dagger) \times (0,T), \\ p^j_{b,q}(0,t) = b^j(t), & t \in (0,T), \\ p^j_{b,q}(a,0) = p^j_0(a), & a \in (0,a_\dagger). \end{cases} \tag{1.6}$$

Here $\{\varphi_j\}_{j\in\mathbf{N}} \subset L^2(\Omega)$ is an orthonormal basis and satisfies (2) in Appendix 2 and

$$f(a,t) - \mu(a,t)q(a,t) = \sum_{j=0}^{\infty} v^j(a,t)\varphi_j,$$

in $L^2(\Omega)$, a.e. $(a,t) \in (0,a_\dagger) \times (0,T)$,

$$b(t) = \sum_{j=0}^{\infty} b^j(t)\varphi_j, \quad \text{in } L^2(\Omega), \text{ a.e. } t \in (0,T),$$

$$p_0(a) = \sum_{j=0}^{\infty} p^j_0(t)\varphi_j, \quad \text{in } L^2(\Omega), \text{ a.e. } a \in (0,a_\dagger).$$

Multiplying $(1.6)_1$ by $p^j_{b,q}$ we obtain that

$$\frac{1}{2}D(|p^j_{b,q}(a,t)|^2) \le |v^j(a,t)| \cdot |p^j_{b,q}(a,t)|$$

$$\le \frac{1}{2}|p^j_{b,q}(a,t)|^2 + \frac{1}{2}|v^j(a,t)|^2,$$

and integrating on $(0, a_\dagger) \times (0, t)$ we get

$$\int_0^{a_\dagger} |p^j_{b,q}(a,t)|^2 da \le \int_0^{a_\dagger} |p^j_0(a)|^2 da + \int_0^t |b^j(s)|^2 ds$$

$$+ \int_0^t \int_0^{a_\dagger} |v^j(a,s)|^2 da\, ds + \int_0^t \int_0^{a_\dagger} |p^j_{b,q}(a,s)|^2 da\, ds.$$

By Bellman's lemma it follows that

$$\|p^j_{b,q}(t)\|^2_{L^2(0,a_\dagger)} \le e^t \left(\|p^j_0\|^2_{L^2(0,a_\dagger)} + \|b^j\|^2_{L^2(0,T)} + \|v^j\|^2_{L^2((0,a_\dagger)\times(0,T))} \right),$$

$\forall t \in (0,T)$ and as a consequence

$$\|p^j_{b,q}\|^2_{L^2((0,a_\dagger)\times(0,T))} \le e^T \left(\|p^j_0\|^2_{L^2(0,a_\dagger)} + \|b^j\|^2_{L^2(0,T)} + \|v_j\|^2_{L^2((0,a_\dagger)\times(0,T))} \right),$$

$\forall j \in \mathbf{N}$. In conclusion

$$p_{b,q} \in L^2(Q_T)$$

and

$$\|p_{b,q}\|^2_{L^2(Q_T)} = \sum_{j=0}^{\infty} \|p^j_{b,q}\|^2_{L^2((0,a_\dagger)\times(0,T))}$$

$$\le e^T \left(\|p_0\|^2_{L^2((0,a_\dagger)\times\Omega)} + \|b\|^2_{L^2((0,T)\times\Omega)} + \|f - \mu q\|^2_{L^2(Q_T)} \right)$$

(as a consequence of Parseval's formula).

The function $p_{b,q}$ with these properties is unique because the solution of (1.4)′ is unique.

Using the Banach fixed point theorem we may infer that there exists a unique p_b, a solution of

$$\begin{cases} Dp + \mu p - k\Delta p = f, & (a,t,x) \in Q_T, \\ \dfrac{\partial p}{\partial \nu}(a,t,x) = 0, & (a,t,x) \in \Sigma_T, \\ p(0,t,x) = b(t,x), & (t,x) \in (0,T) \times \Omega, \\ p(0,t,x) = p_0(a,x), & (a,x) \in (0,a_\dagger) \times \Omega. \end{cases} \tag{1.6'}$$

This means that $p_b \in L^2(Q_T)$,

$$p_b \in C(\overline{S}; L^2(\Omega)) \cap AC(S; L^2(\Omega)) \cap L^2(S; H^1(\Omega)) \cap L^2_{loc}(S; H^2(\Omega)),$$

for almost any characteristic line S and satisfies

$$\begin{cases} Dp_b + \mu p_b - k\Delta p_b = f, & \text{a.e. in } Q_T, \\ \dfrac{\partial p_b}{\partial \nu}(a,t,x) = 0, & \text{a.e. in } \Sigma_T, \\ \lim\limits_{\varepsilon \to 0^+} p_b(\varepsilon, t+\varepsilon, \cdot) = b(t,\cdot), & \text{in } L^2(\Omega), \text{ a.e. } t \in (0,T), \\ \lim\limits_{\varepsilon \to 0^+} p_b(a+\varepsilon, \varepsilon, \cdot) = p_0(a,\cdot), & \text{in } L^2(\Omega), \text{ a.e. } a \in (0,a_\dagger). \end{cases}$$

We need to establish the following auxiliary result

LEMMA 4.1.2. *Under the hypotheses of Lemma 4.1.1 and for any* $b_1, b_2 \in L^2((0,T)\times\Omega)$, $0 \le b_1(t,x) \le b_2(t,x)$ *a.e. in* $(0,T)\times\Omega$, *we have that*

$$0 \le p_{b_1}(a,t,x) \le p_{b_2}(a,t,x) \text{ a.e. in } Q_T.$$

Proof. Consider $v = p_{b_1} - p_{b_2}$, which is a solution of

$$\begin{cases} Dv + \mu v - k\Delta v = 0, & (a,t,x) \in Q_T, \\ \dfrac{\partial v}{\partial \nu}(a,t,x) = 0, & (a,t,x) \in \Sigma_T, \\ v(0,t,x) = b_1(t,x) - b_2(t,x), & (t,x) \in (0,T)\times\Omega, \\ v(a,0,x) = 0, & (a,x) \in (0,a_\dagger)\times\Omega. \end{cases}$$

Multiplying the first equation by v^+ and integrating over Q_t we get that

$$\frac{1}{2}\int_0^{a_\dagger}\int_\Omega |v^+(a,t,x)|^2 dx\, da \le \int_0^t \int_\Omega (b_1 - b_2)(s,x) v^+(0,s,x) dx\, ds$$

$\forall t \in [0,T]$ and consequently $v^+(a,t,x) = 0$ a.e. in Q_T and so we get the conclusion of the lemma.

Proof of Lemma 4.1.1 - continued. Denote by

$$L^2_+((0,T)\times\Omega) = \{h \in L^2((0,T)\times\Omega); 0 \le h(t,x) \text{ a.e. in } (0,T)\times\Omega\}$$

and consider the operator $\mathcal{F} : L^2((0,T)\times\Omega) \to L^2((0,T)\times\Omega)$ defined by

$$(\mathcal{F}b)(t,x) = \int_0^{a_\dagger} \beta(a,t,x) p_b(a,t,x) da \quad \text{a.e. } (t,x) \in (0,T)\times\Omega.$$

For any $h_1, h_2 \in L^2((0,T)\times\Omega)$, we denote by $w = p_{h_1} - p_{h_2}$; w is the solution of

$$\begin{cases} Dw + \mu w - k\Delta w = 0, & (a,t,x) \in Q_T, \\ \dfrac{\partial w}{\partial \nu}(a,t,x) = 0, & (a,t,x) \in \Sigma_T, \\ w(0,t,x) = h_1(t,x) - h_2(t,x), & (t,x) \in (0,T)\times\Omega, \\ w(a,0,x) = 0, & (a,x) \in (0,a_\dagger)\times\Omega. \end{cases}$$

Multiplying the first equation by w and integrating over $(0,a_\dagger)\times(0,t)\times\Omega$ we get

$$\|w(t)\|^2_{L^2((0,a_\dagger)\times\Omega)} \le \int_0^t \|h_1(s) - h_2(s)\|^2_{L^2(\Omega)} ds$$

(we have used the Fubini theorem) and consequently

$$\int_0^T e^{-\lambda t}\|w(t)\|^2_{L^2((0,a_\dagger)\times\Omega)}dt = \int_0^T \int_0^t e^{-\lambda t}\|h_1(s) - h_2(s)\|^2_{L^2(\Omega)}ds\ dt$$

$$= \int_0^T \int_s^T e^{-\lambda t}\|h_1(s) - h_2(s)\|^2_{L^2(\Omega)}dt\ ds.$$

This implies

$$\int_0^T e^{-\lambda t}\|w(t)\|^2_{L^2((0,a_\dagger)\times\Omega)}dt \le \frac{1}{\lambda}\int_0^T e^{-\lambda t}\|h_1(t) - h_2(t)\|^2_{L^2(\Omega)}dt, \tag{1.7}$$

for any $\lambda \in (0,+\infty)$.

Consider on $L^2((0,T)\times\Omega)$ the norm

$$\|h\| = \left(\int_0^T e^{-\lambda t}\|h(t)\|^2_{L^2(\Omega)}dt\right)^{\frac{1}{2}}, \quad \forall h \in L^2((0,T)\times\Omega),$$

which is equivalent to the usual norm (the constant $\lambda > 0$ will be precised later).

We obtain

$$\|\mathcal{F}h_1 - \mathcal{F}h_2\|^2 = \int_0^T e^{-\lambda t}\|\int_0^{a_\dagger} \beta(a,t,x)(p_{h_1} - p_{h_2})(a,t,x)da\|^2_{L^2(\Omega)}dt$$

$$\le a_\dagger\|\beta\|^2_{L^\infty(Q_T)}\int_0^T e^{-\lambda t}\|w(t)\|^2_{L^2((0,a_\dagger)\times\Omega)}dt$$

and by (1.7) we get

$$\|\mathcal{F}h_1 - \mathcal{F}h_2\|^2 = \|\beta\|^2_{L^\infty(Q_T)}\frac{a_\dagger}{\lambda}\|h_1 - h_2\|^2.$$

It is now obvious that for any $\lambda > a_\dagger \|\beta\|^2_{L^\infty(Q_T)}$, $\mathcal{F}$ is a contraction on $(L^2((0,T)\times\Omega), d)$, where d is the metric defined by

$$d(h_1,h_2) = \|h_1 - h_2\|, \quad \forall h_1, h_2 \in L^2((0,T)\times\Omega).$$

Banach's fixed point theorem allows us to conclude that there exists a unique $b \in L^2((0,T)\times\Omega)$ such that $b = \mathcal{F}b$. Since $\mathcal{F}(L^2_+((0,T)\times\Omega)) \subset (L^2_+((0,T)\times\Omega))$, and $L^2_+((0,T)\times\Omega)$ is closed in $(L^2((0,T)\times\Omega), d)$, then we conclude that this b belongs to $L^2_+((0,T)\times\Omega)$ and as a consequence of Lemma A2.7 we obtain that p, the solution of (1.1), is nonnegative.

Now, for the second part of Lemma 4.1.1: If $\beta_i, \mu_i, f_i, p_{0i}$ $(i \in \{1,2\})$ satisfy the hypotheses in Lemma 4.1.1, then we define the operators

$$\mathcal{F}_i : L^2((0,T)\times\Omega) \to L^2((0,T)\times\Omega),$$

$$(\mathcal{F}_i b)(t,x) = \int_0^{a_\dagger} \beta(a,t,x) p_{i,b}(a,t,x)da \text{ a.e. } (t,x) \in (0,T)\times\Omega,$$

where $p_{i,b}$ are the solutions of $(1.6)'$ corresponding to $(\beta_i, \mu_i, f_i, p_{0i})$, respectively.

By Lemma A2.7 we conclude that for any $b \in L^2_+((0,T)\times\Omega))$:

$$p_{1,b}(a,t,x) \le p_{2,b}(a,t,x) \quad \text{a.e. in } Q_T,$$

which implies

$$(\mathcal{F}_1 b)(t,x) \le (\mathcal{F}_2 b)(t,x) \quad \text{a.e. in } (0,T)\times\Omega.$$

We have already proved that there exists a unique fixed point b_2 of $\mathcal{F}_2$ in $L^2_+((0,T)\times\Omega)$. Since, the set

$$\mathcal{C} = \{b \in L^2((0,T)\times\Omega); 0 \le b(t,x) \le b_2(t,x) \text{ a.e. in } (0,T)\times\Omega\}$$

is closed ($(\mathcal{C}, d)$ is a complete metric space) and so $\mathcal{F}_1$ has a unique fixed point $b_1 \in \mathcal{C}$ (we have used Banach's fixed point theorem). So,

$$0 \le b_1(t,x) \le b_2(t,x) \text{ a.e. in } (0,T)\times\Omega$$

and the conclusion follows by Lemma 4.1.2.

We can formulate now, the main result of this section:

THEOREM 4.1.3. *System (1.1) has a unique solution. The solution is nonnegative.*

Proof. For any $N \in \mathbf{N}^*$ we define

$$\mu^N(a,t,x) = \min\{\mu(a,t,x), N\}, \quad (a,t,x) \in Q_T.$$

μ^N satisfies assumption $(\mathbf{A_2})'$ and $\{\mu^N\}$ is an increasing sequence.

Denote by p_N the solution of (1.1), corresponding to $\mu := \mu^N$. Applying Lemma 4.1.1 we obtain that

$$p_{N_1}(a,t,x) \geq p_{N_2}(a,t,x) \quad \text{a.e. in } Q_T, \tag{1.8}$$

for any $N_1, N_2 \in \mathbf{N}^*, N_1 \leq N_2$. This relation and Beppo-Levi's theorem imply that

$$p_N(a,t,x) \to p(a,t,x) \quad \text{a.e. } (a,t,x) \in Q_T, \tag{1.9}$$

(as $N \to +\infty$) and that $p(a,t,x) \geq 0$ a.e. in $Q_T, p \in L^2(Q_T)$ $(0 \leq p(a,t,x) \leq p_1(a,t,x)$ a.e. in Q_T; here p_1 is the solution of (1.1) corresponding to $\mu := \mu^1)$.

By (1.8) and (1.9) we get via Lebesque's theorem that

$$\int_0^{a_\dagger} \beta(a,t,x)p_N(a,t,x)da \to \int_0^{a_\dagger} \beta(a,t,x)p(a,t,x)da \tag{1.10}$$

a.e. in $(0,T) \times \Omega$, as $N \to +\infty$.

We shall prove that p is the unique solution to (1.1).

Since $\mu \in L^\infty_{loc}([0,a_\dagger) \times [0,T] \times \overline{\Omega})$, we obtain that for any $\varepsilon > 0$, small enough, there exists $N_\varepsilon \in \mathbf{N}^*$ such that

$$\mu(a,t,x) = \mu^N(a,t,x) \quad \text{a.e. in } (0,a_\dagger - \varepsilon) \times (0,T) \times \Omega, \tag{1.11}$$

for any $N \geq N_\varepsilon$.

Using now the continuous dependence of the solution to the linear parabolic equation on the initial datum, (1.10) and (1.11), we conclude that

$$p \in C(S_1; L^2(\Omega)) \cap AC(S; L^2(\Omega)) \cap L^2_{loc}(S; H^2(\Omega))$$

for almost any characteristic lines S and S_1, given by

$$S = \{(a,t) \in (0,a_\dagger) \times (0,T); a - t = a_0 - t_0\}$$

$$= \{(a_0 + s, t_0 + s); s \in (0,\alpha)\}, \ (a_0,t_0) \in \{0\} \times (0,T) \cup (0,a_\dagger) \times \{0\},$$

$$S_1 = \begin{cases} \{(a,t) \in [0,a_\dagger) \times [0,T); a - t = a_0 - t_0\} & \\ \quad = \{(a_0 + s, t_0 + s); s \in (0,\alpha)\} & \text{if } \alpha = a_\dagger - a_0, \\ \overline{S} = \{(a,t) \in [0,a_\dagger] \times [0,T]; a - t = a_0 - t_0\}, & \text{elsewhere.} \end{cases}$$

On the other hand p satisfies (1.2). It remains to prove that

$$p \in C(\overline{S}; L^2(\Omega)) \cap L^2(S; H^1(\Omega)),$$

for

$$\overline{S} = \{(a,t) \in [0,a_\dagger] \times [0,T]; a - t = a_0 - t_0\},$$

if $\alpha = a_\dagger - a_0$. For this it suffices to show that

$$\lim_{s \nearrow \alpha} p(a_0 + s, t_0 + s) = 0 \quad \text{in } L^2(\Omega),$$

if $\alpha = a_\dagger - a_0$.

Multiplying $(1.2)_1$ by p and integrating over Ω we get that

$$\frac{d}{ds}\|p(a_0+s,t_0+s)\|^2_{L^2(\Omega)} + \mu_0(a_0+s,t_0+s)\|p(a_0+s,t_0+s)\|^2_{L^2(\Omega)}$$

$$\leq \|f(a_0+s,t_0+s)\|_{L^2(\Omega)} \cdot \|p(a_0+s,t_0+s)\|_{L^2(\Omega)}, \quad \forall s \in (0,\alpha)$$

and

$$\frac{d}{ds}\{e^{\int_0^s \mu_0(a_0+\tau,t_0+\tau)d\tau}\|p(a_0+s,t_0+s)\|^2_{L^2(\Omega)}\}$$

$$\leq e^{\int_0^s \mu_0(a_0+\tau,t_0+\tau)d\tau}\|f(a_0+s,t_0+s)\|_{L^2(\Omega)} \cdot \|p(a_0+s,t_0+s)\|_{L^2(\Omega)}.$$

Integrating the last relation we get that

$$\|p(a_0+s,t_0+s)\|^2_{L^2(\Omega)} \leq e^{-\int_0^s \mu_0(a_0+\tau,t_0+\tau)d\tau}\|p(a_0,t_0)\|^2_{L^2(\Omega)}$$

$$+ \int_0^s e^{-\int_u^s \mu_0(a_0+\tau,t_0+\tau)d\tau}\|f(a_0+u,t_0+u)\|_{L^2(\Omega)}$$

$$\cdot\|p(a_0+u,t_0+u)\|_{L^2(\Omega)}du, \tag{1.12}$$

$\forall s \in [0,\alpha]$, a.e. $s \in (0,\alpha)$. For $s \nearrow \alpha$, we obtain (by (1.12)) that

$$\lim_{s \nearrow \alpha} p(a_0 + s, t_0 + s) = 0 \quad \text{in } L^2(\Omega),$$

if $\alpha = a_\dagger - a_0$ (because $\int_0^{a_\dagger} \mu_0(a_0+s,t_0+s)ds = +\infty$). We may infer that $p \in C(\overline{S}; L^2(\Omega))$.

Multiplying again $(1.2)_1$ by p and integrating on $S \times \Omega$ we get

$$k\int_S\int_\Omega |\nabla p(a_0+s,t_0+s,x)|^2 dx\, dl \leq \frac{1}{2}\|p(a_0,t_0)\|^2_{L^2(\Omega)}$$

and it follows that $p \in L^2(S; H^1(\Omega))$. So, p is a solution of (1.1) (and p is nonnegative).

REMARK 4.1.1. *If $f(a,t,x) > 0$ a.e. in Q_T, then we may conclude that*

$$p(a,t,x) > 0 \quad \text{a.e. in } Q_T.$$

This follows by Lemma A2.7.

The following comparison result holds:

THEOREM 4.1.4. *If p_i $(i \in \{1,2\})$ are the solutions of (1.1), corresponding to $(\beta_i, \mu_i, f_i, p_{0i})$, respectively and if $(\beta_i, \mu_i, f_i, p_{0i})$ satisfy* **(A1), (A2), (A3)** *and*

$$\beta_1 \le \beta_2,\ \mu_1 \ge \mu_2,\ f_1 \le f_2,\ p_{01} \le p_{02},$$

then

$$0 \le p_1(a,t,x) \le p_2(a,t,x) \quad \text{a.e. in } Q_T.$$

Proof. Let p_{iN} be the solution of (1.1), corresponding to $\beta := \beta_i$, $f := f_i$, $p_0 := p_{0i}$ and $\mu := \mu_i^N$, where

$$\mu_i^N = \min\{\mu_i, N\}.$$

By Lemma 4.1.1 we get that

$$0 \le p_{1N}(a,t,x) \le p_{2N}(a,t,x) \text{ a.e. in } Q_T$$

and passing here to the limit we obtain the conclusion of the theorem.

THEOREM 4.1.5. *If $\{f_N\}$ satisfy* **(A3)** $(N \in \mathbf{N})$ *and*

$$f_N \to f, \quad \text{in } L^2(Q_T)$$

(as $N \to +\infty$), then

$$p_{f_N} \to p, \quad \text{in } L^\infty(0,T; L^2((0,a_\dagger)\times\Omega)),$$

where p_{f_N} is the solution of (1.1) corresponding to $f := f_N$ and p is the solution of (1.1) corresponding to f.

Proof. Since f_N satisfy **(A3)**, then f (the limit of f_N in $L^2(Q_T)$) also satisfies **(A3)**. Denote by $w_N = p_{f_N} - p$; w_N is the solution to

$$\begin{cases} Dw + \mu w - k\Delta w = f_N - f, & \text{in } Q_T, \\ \dfrac{\partial w}{\partial \nu}(a,t,x) = 0, & \text{on } \Sigma_T, \\ w(0,t,x) = \displaystyle\int_0^{a_\dagger} \beta(a,t,x)w(a,t,x)da, & \text{in } (0,T)\times\Omega, \\ w(a,0,x) = 0, & \text{in } (0,a_\dagger)\times\Omega. \end{cases} \tag{1.13}$$

Multiplying the first equation in (1.13) by w_N and integrating on $(0, a_\dagger) \times (0,t) \times \Omega$ $(t \in (0,T))$ we obtain

$$\frac{1}{2}\|w_N(t)\|^2_{L^2((0,a_\dagger)\times\Omega)} \le \frac{1}{2}\int_0^t \int_\Omega |w_N(0,s,x)|^2 dx\ ds$$

$$+\int_0^t \int_0^{a_\dagger} \int_\Omega (f_N - f)(a,s,x) w_N(a,s,x) dx\ da\ ds$$

$$\le \frac{1}{2} a_\dagger \|\beta\|^2_{L^\infty(Q_T)} \int_0^t \|w_N(s)\|^2_{L^2((0,a_\dagger)\times\Omega)} ds$$

$$+\frac{1}{2}\|f_N - f\|^2_{L^2(Q_T)} + \frac{1}{2}\int_0^t \|w_N(s)\|^2_{L^2((0,a_\dagger)\times\Omega)} ds,$$

$\forall t \in [0,T]$. By Bellman's inequality we get

$$\|w_N(t)\|_{L^2((0,a_\dagger)\times\Omega)} \le e^{t(1+a_\dagger\|\beta\|^2_{L^\infty(Q_T)})} \|f_N - f\|_{L^2(Q_T)},$$

$\forall t \in [0,T]$ and as a consequence

$$w_N \to 0 \quad \text{in } L^\infty(0,T; L^2((0,a_\dagger)\times\Omega))$$

and

$$p_{f_N} \to p \quad \text{in } L^\infty(0,T; L^2((0,a_\dagger)\times\Omega)).$$

We shall end this section with a discussion concerning system (1.1), with β, μ depending only on age and with $f \equiv 0$. So, we deal with the system

$$\begin{cases} Dp + \mu(a)p - k\Delta p = 0, & (a,t,x) \in Q_T, \\ \dfrac{\partial p}{\partial \nu}(a,t,x) = 0, & (a,t,x) \in \Sigma_T, \\ p(0,t,x) = \displaystyle\int_0^{a_\dagger} \beta(a) p(a,t,x) da, & (t,x) \in (0,T)\times\Omega, \\ p(a,0,x) = p_0(a,x), & (a,x) \in (0,a_\dagger)\times\Omega. \end{cases} \tag{1.14}$$

Here we assume that

(Ai1) $\beta \in L^\infty(0,a_\dagger), \quad \beta(a) \ge 0$ a.e. in $(0,a_\dagger)$,

$$\textbf{(Ai2)} \qquad \begin{cases} \mu \in L^\infty_{loc}([0, a_\dagger)), \\ \mu(a) \geq 0, & \text{a.e. in } (0, a_\dagger), \\ \displaystyle\int_0^{a_\dagger} \mu_0(a)da = +\infty, \end{cases}$$

(Ai3) $\quad p_0 \in L^2((0, a_\dagger) \times \Omega),\ p_0(a, x) \geq 0$ a.e. in $(0, a_\dagger) \times \Omega$.

If we apply the result from the beginning of the section to our problem, we conclude that the solution p of (1.14) satisfies

$$p(a, t) = \sum_{j=0}^{\infty} p^j(a, t)\varphi_j \quad \text{in } L^2(\Omega), \text{ a.e. in } (0, a_\dagger) \times (0, T),$$

where

$$\begin{cases} Dp^j + \mu p^j + \lambda_j k p^j = 0, & (a, t) \in (0, a_\dagger) \times (0, T), \\ p^j(0, t) = \displaystyle\int_0^{a_\dagger} \beta(a)p^j(a, t)da, & t \in (0, T), \\ p^j(a, 0) = p_0^j(a), & a \in (0, a_\dagger). \end{cases} \tag{1.15}$$

Here

$$p_0(a) = \sum_{j=0}^{\infty} p_0^j(a)\varphi_j \quad \text{in } L^2(\Omega), \text{ a.e. } a \in (0, a_\dagger).$$

Using now Parseval's formula we get that

$$\|p_0(a)\|^2_{L^2(\Omega)} = \sum_{j=0}^{\infty} (p_0^j)^2(a)$$

and so

$$\|p_0\|^2_{L^2((0,a_\dagger)\times\Omega)} = \sum_{j=0}^{\infty} \|p_0^j\|^2_{L^2(0,a_\dagger)}.$$

We have shown in Section 2.3, that p^j (the unique solution of (1.15) is continuous on

$$\mathcal{C}_1 = \{(a, t) \in [0, a_\dagger] \times [0, T]; a < t\}$$

$\forall j \in \mathbf{N}$. Let us prove that $p \in C(\mathcal{C}_1; L^2(\Omega))$. Indeed, for any $(\widetilde{a}, \widetilde{t}) \in \mathcal{C}_1$, we have

$$\|p(a, t) - p(\widetilde{a}, \widetilde{t})\|^2_{L^2(\Omega)} = \sum_{j=0}^{\infty} |p^j(a, t) - p^j(\widetilde{a}, \widetilde{t})|^2$$

$$= \sum_{j=0}^{N} |p^j(a,t) - p^j(\widetilde{a},\widetilde{t})|^2 + \sum_{j=N+1}^{\infty} |p^j(a,t) - p^j(\widetilde{a},\widetilde{t})|^2$$

$$\leq \sum_{j=0}^{N} |p^j(a,t) - p^j(\widetilde{a},\widetilde{t})|^2$$

$$+ \sum_{j=N+1}^{\infty} e^{ta_\dagger \|\beta\|^2_{L^\infty(0,a_\dagger)}} \cdot \|p_0^j\|^2_{L^2(0,a_\dagger)}$$

$$\leq \sum_{j=0}^{N} |p^j(a,t) - p^j(\widetilde{a},\widetilde{t})|^2 + C_2 \sum_{j=N+1}^{\infty} \|p_0^j\|^2_{L^2(0,a_\dagger)},$$

where C_2 is a positive constant.

Since $\sum_{j=0}^{\infty} \|p_0^j\|^2_{L^2(0,a_\dagger)}$ is convergent in $\mathbf{R}$, we conclude that for any $\varepsilon > 0$, there exists $N_\varepsilon \in \mathbf{N}$, such that

$$C_2 \sum_{j=N_\varepsilon+1}^{\infty} \|p_0^j\|^2_{L^2(0,a_\dagger)} < \varepsilon.$$

Using this and the continuouity of p^j, $j \in \{0, 1, ..., N_\varepsilon\}$ (in $(\widetilde{a},\widetilde{t})$) we deduce that $p \in C(\mathcal{C}_1; L^2(\Omega))$.

4.2. A general nonlinear model

This section concerns a quite general nonlinear model with diffusion

$$\begin{cases} Dp - k\Delta p + \mu(a,t,x,P(t,x))p = f & \text{in } Q_T, \\ \dfrac{\partial p}{\partial \nu}(a,t,x) = 0, & \text{on } \Sigma_T, \\ p(0,t,x) = \displaystyle\int_0^{a_\dagger} \beta(a,t,x,P(t,x))p(a,t,x)da, & \text{in } (0,T) \times \Omega, \\ p(a,0,x) = p_0(a,x), & \text{in } (0,a_\dagger) \times \Omega, \\ P(t,x) = \displaystyle\int_0^{a_\dagger} p(a,t,x)da, & \text{in } (0,T) \times \Omega. \end{cases} \tag{2.1}$$

We shall work under the following hypotheses:

(An1) $\forall s \in \mathbf{R}^+$ the functions $\beta(\cdot,\cdot,\cdot,s)$ and $\mu(\cdot,\cdot,,\cdot,s)$ belong to $L^\infty(\overline{Q}_T)$ and $L^\infty_{loc}([0,a_\dagger) \times [0,T] \times \overline{\Omega})$, respectively;

β and μ are locally Lipschitz functions with respect to the fourth variable, i.e.,
For any $M > 0$, there exists $L(M) > 0$ such that
$|\beta(a,t,x,s_1) - \beta(a,t,x,s_2)| \leq L(M) \cdot |s_1 - s_2|$
$|\mu(a,t,x,s_1) - \mu(a,t,x,s_2)| \leq L(M) \cdot |s_1 - s_2|, \quad \forall s_1, s_2 \in [0, M]$,
a.e. $(a,t,x) \in Q_T$;

(An2) $\beta(a,t,x,s),\ \mu(a,t,x,s) \geq 0$ a.e. in $Q_T \times (0,+\infty)$,
$\beta(a,t,x,\cdot)$ is nonincreasing a.e. $(a,t,x) \in Q_T$ and
$\mu(a,t,x,\cdot)$ is nondecreasing a.e. $(a,t,x) \in Q_T$;

(An3) $\mu(a,t,x,P) \geq \mu_0(a,t) \geq 0$ a.e. in $Q_T \times \mathbf{R}^+$,
$\int_0^{a_\dagger} \mu_0(a, t - a_\dagger + a)da = +\infty$ a.e. $t \in (0,T)$
(here μ_0 is extended by zero on $(0,a_\dagger) \times (-\infty, 0)$);

(An4) $p_0 \in L^\infty((0,a_\dagger) \times \Omega),\ p_0(a,x) \geq 0$ a.e. $a \in (0,a_\dagger) \times \Omega$
$f \in L^\infty(Q_T),\ f(a,t,x) \geq 0$ a.e. $(a,t,x) \in Q_T$.

By *a solution* to (2.1) we mean a solution of (1.1), where $\beta(a,t,x) := \beta(a,t,x,P(t,x))$, $\mu(a,t,x) := \mu(a,t,x,P(t,x))$ and

$$P(t,x) = \int_0^{a_\dagger} p(a,t,x)da \quad \text{a.e. in } (0,T) \times \Omega.$$

THEOREM 4.2.1. *System (2.1) has a unique solution. The solution is nonnegative.*

Proof. If (2.1) admits a solution, this solution must be nonnegative because the solution to the linear model (1.1) in Section 4.1 is nonnegative.

Denote bu $\Lambda : L^2_+(Q_T) \to L^2_+(Q_T)$, the application given by

$$\Lambda h = p^h,$$

where p^h is the solution of

$$\begin{cases} Dp - k\Delta p + \mu(a,t,x,H(t,x))p = f & \text{in } Q_T, \\ \dfrac{\partial p}{\partial \nu}(a,t,x) = 0, & \text{on } \Sigma_T, \\ p(0,t,x) = \displaystyle\int_0^{a_\dagger} \beta(a,t,x,H(t,x))p(a,t,x)da, & \text{in } (0,T) \times \Omega, \\ p(a,0,x) = p_0(a,x), & \text{in } (0,a_\dagger) \times \Omega, \end{cases}$$

where $H(t,x) = \int_0^{a_\dagger} h(a,t,x)da$ a.e. in $(0,T)\times\Omega$.

The application Λ is well defined (by Theorem 4.1.3). The comparison Theorem 4.1.4 implies that for any $h \in L^2_+(Q_T)$

$$0 \le p^h(a,t,x) \le \overline{p}(a,t,x) \quad \text{a.e. in } Q_T,$$

where $\overline{p}$ is the solution of (2.1) corresponding to a null mortality rate and to a maximal fertility rate ($= \|\beta\|_{L^\infty(Q_T\times\mathbf{R}^+)}$), to $f := \|f\|_{L^\infty(Q_T)}$, $p_0 := \|p_0\|_{L^\infty((0,a_\dagger)\times\Omega)}$;
$\overline{p}$ does not depend on $x\in\overline{\Omega}$ and using a remark in Section 2.1, $\overline{p}\in L^\infty(Q_T)$.

For any $h_1, h_2 \in L^2_+(Q_T)$, $0 \le h_i(a,t,x) \le \overline{p}(a,t,x)$ ($i\in\{1,2\}$) a.e. in Q_T, we define

$$H_i(t,x) = \int_0^{a_\dagger} h_i(a,t,x)da \quad \text{a.e. } (t,x)\in(0,T)\times\Omega,$$

$i\in\{1,2\}$ and $w = \Lambda h_1 - \Lambda h_2$; w is the solution of

$$\begin{cases} Dw - k\Delta w + \mu(a,t,x,H_1(t,x))w & \\ = (\mu(a,t,x,H_2(t,x)) - \mu(a,t,x,H_1(t,x)))p^{h_2} & (a,t,x)\in Q_T, \\ \dfrac{\partial w}{\partial\nu}(a,t,x) = 0, & (a,t,x)\in\Sigma_T, \\ w(0,t,x) = \displaystyle\int_0^{a_\dagger}\beta(a,t,x)da, & (t,x)\in(0,T)\times\Omega, \\ w(a,0,x) = 0, & (a,x)\in(0,a_\dagger)\times\Omega. \end{cases}$$

Multiplying by w and integrating over $(0,a_\dagger)\times(0,t)\times\Omega$ we get that

$$\|(\Lambda h_1 - \Lambda h_2)(t)\|^2_{L^2((0,a_\dagger)\times\Omega)} \le C\int_0^t \|(h_1-h_2)(s)\|^2_{L^2((0,a_\dagger)\times\Omega)}ds,$$

where $C>0$ is a constant which depends on the Lipschitz constants for β and μ, with respect to P.

Consider now the set

$$\mathcal{C} = \{h\in L^2(Q_T); 0\le h(a,t,x)\le\overline{p}(a,t,x) \text{ a.e. in } Q_T\},$$

which is closed in $L^2(Q_T)$ and $\Lambda(\mathcal{C})\subset\mathcal{C}$, because $0\le(\Lambda h)(a,t,x)\le\overline{p}(a,t,x)$ a.e. in Q_T, for any $h\in\mathcal{C}$.

On the other hand, define the metric d on $L^2(Q_T)$:

$$d(h_1,h_2) = \left(\int_0^T \|h_1(t) - h_2(t)\|^2_{L^2((0,a_\dagger)\times\Omega)} e^{-2Ct} dt \right)^{\frac{1}{2}},$$

which is equivalent to the usual metric on $L^2(Q_T)$.

With respect to this metric, Λ is a contraction. Indeed, it is obvious that

$$\int_0^T e^{-2Ct} \|(\Lambda h_1 - \Lambda h_2)(t)\|^2_{L^2((0,a_\dagger)\times\Omega)} dt$$

$$\leq C \int_0^T e^{-2Ct} \int_0^t \|(h_1 - h_2)(s)\|^2_{L^2((0,a_\dagger)\times\Omega)} ds\ dt$$

$$= \int_0^T \|(h_1 - h_2)(s)\|^2_{L^2((0,a_\dagger)\times\Omega)} \int_s^T Ce^{-2Ct} dt\ ds$$

and as a consequence

$$d(\Lambda h_1, \Lambda h_2) \leq \frac{1}{2} d(h_1,h_2).$$

So, Λ is a contraction on the complete metric space $(\mathcal{C}, d)$ and via Banach's fixed point theorem we conclude the existence of a unique fixed point for Λ.

We conclude the existence of a nonnegative solution p. Any other solution q should satisfy

$$0 \leq q(a,t,x) \leq \overline{p}(a,t,x) \quad \text{a.e. in } Q_T,$$

because of the comparison result for the linear model (see Section 4.1). So $\Lambda q = q$ and $q \in \mathcal{C}$. The conclusion is that $p \equiv q$.

As a consequence of the comparison result for the linear model, in the previous section (Theorem 4.1.4), the following result holds:

THEOREM 4.2.2. *If p_i $(i \in \{1,2\})$ are the solutions of (2.1), corresponding to $(\beta_i, \mu_i, f_i, p_{0i})$ and if $(\beta_i, \mu_i, f_i, p_{0i})$ satisfy* **(An1)-(An4)** *and*

$$\beta_1 \leq \beta_2,\ \mu_1 \geq \mu_2,\ f_1 \leq f_2,\ p_{01} \leq p_{02},$$

then

$$0 \leq p_1(a,t,x) \leq p_2(a,t,x) \quad \text{a.e. in } Q_T.$$

We leave the proof of this result to the reader.

An important particular situation is that of fertility rate β depending only on age a and time t and of the mortality rate μ with the structure

$$\mu(a,t,P) = \mu_0(a,t) + \mathcal{M}(P).$$

Here β and μ_0 are the natural fertility and mortality rates, respectively; $\mathcal{M}(P)$ is an additional mortality rate due to the limited resources of the habitat.

We wish to analyze the structure of the solution to the autonomous system

$$\begin{cases} Dp + \mu_0(a,t)p + \mathcal{M}(P(t,x))p - k\Delta p = 0 & \text{in } Q_T, \\ \dfrac{\partial p}{\partial \nu}(a,t,x) = 0, & \text{on } \Sigma_T, \\ p(0,t,x) = \displaystyle\int_0^{a_\dagger} \beta(a,t)p(a,t,x)da & \text{in } (0,T)\times\Omega, \\ p(a,0,x) = \overline{p}_0(a)h_0(x) & \text{in } (0,a_\dagger)\times\Omega, \end{cases} \tag{2.2}$$

where

$$P(t,x) = \int_0^{a_\dagger} p(a,t,x)da \qquad \text{in } (0,T)\times\Omega.$$

We assume that **(Ans1)-(Ans3)** in Section 2.2 hold and that

$$(\mathbf{Ans4})' \qquad \begin{cases} \overline{p}_0 \in L^2(0,a_\dagger)\ ,\ \overline{p}_0(a) \geq 0 \quad \text{a.e. in } (0,a_\dagger), \\ h_0 \in L^2(\Omega)\ ,\ h_0(x) \geq 0 \quad \text{a.e. in } \Omega. \end{cases}$$

Problem (2.2) belongs to the class of "separable" models considered in Section 2.2. We leave to the reader the demonstration of the fact that the unique solution p of (2.2) satisfies

$$p(a,t,x) = \tilde{p}(a,t)h(t,x) \quad \text{a.e. in } Q_T,$$

where $\tilde{p}$ is the solution of

$$\begin{cases} D\tilde{p} + \mu_0(a,t)\tilde{p} = 0, & (a,t)\in(0,a_\dagger)\times(0,T), \\ \tilde{p}(0,t) = \displaystyle\int_0^{a_\dagger} \beta(a,t)\tilde{p}(a,t)da, & t\in(0,T), \\ \tilde{p}(a,0) = \overline{p}_0(a), & a\in(0,a_\dagger), \end{cases}$$

and

$$\begin{cases} h_t(t,x) + \mathcal{M}(P_0(t)h(t,x))h(t,x) - k\Delta h(t,x) = 0, & \text{in } (0,T)\times\Omega, \\ \dfrac{\partial h}{\partial \nu}(t,x) = 0, & \text{on } (0,T)\times\partial\Omega, \\ h(0,x) = h_0(x) & \text{in } \Omega. \end{cases}$$

Here

$$P_0(t) = \int_0^{a_\dagger} \tilde{p}(a,t)da, \quad \text{a.e. } t \in (0,T).$$

4.3. Asymptotic behaviour of the solution

We analyze here the asymptotic behaviour of the solution to the system

$$\begin{cases} Dp + \mu(a)p - k\Delta p = 0, & (a,t,x) \in Q, \\ \dfrac{\partial p}{\partial \nu}(a,t,x) = 0, & (a,t,x) \in \Sigma, \\ p(0,t,x) = \displaystyle\int_0^{a_\dagger} \beta(a)p(a,t,x)da, & (t,x) \in (0,\infty)\times\Omega, \\ p(a,0,x) = p_0(a,x), & (a,x) \in (0,a_\dagger)\times\Omega, \end{cases} \tag{3.1}$$

where $Q = (0,a_\dagger)\times(0,+\infty)\times\Omega$ and $\Sigma = (0,a_\dagger)\times(0,+\infty)\times\partial\Omega$. This is a linear model with time and space independent vital rates.

Assume that β, μ and p_0 satisfy **(Ai1)**, **(Ai2)**, **(Ai3)** in Section 4.1.

By *a solution* to (3.1) we mean a function p which is a solution to (1.14) (Section 4.1) for any $T \in (0,+\infty)$. In Section 4.1 we have established that (3.1) has a unique solution p and this solution belongs to $C(\widetilde{Q}; L^2(\Omega))$, where

$$\widetilde{Q} = \{(a,t) \in [0,a_\dagger]\times[0,+\infty); a < t\}.$$

This allows us to infer that

$$p \in C((a_\dagger,+\infty); L^2((0,a_\dagger)\times\Omega)).$$

We have established (in Section 4.1) that

$$p(a,t) = \sum_{j=0}^{\infty} p^j(a,t)\varphi_j \quad \text{in } L^2(\Omega),$$

a.e. $(a,t) \in (0,a_\dagger) \times (0,+\infty)$, where p^j satisfies (1.15) (Section 4.1). Here $\{\varphi_j\}_{j\in\mathbf{N}}$ is the orthonormal basis on $L^2(\Omega)$ (in Section 4.1; see also Appendix 2).

Consider

$$R = \int_0^{a_\dagger} \beta(a) e^{-\int_0^a \mu(s)ds} da.$$

THEOREM 4.3.1. *Assume that $R < 1$. Then*

$$\lim_{t\to+\infty} \|p(t)\|_{L^2((0,a_\dagger)\times\Omega)} = 0.$$

Proof. By Remark 2.3.2 we get that

$$\lim_{t\to+\infty} \|p^j(t)\|_{L^\infty(0,a_\dagger)} = 0,$$

for any $j \in \mathbf{N}$ and consequently

$$\lim_{t\to+\infty} \|p^j(t)\|_{L^2(0,a_\dagger)} = 0, \quad \forall j \in \mathbf{N}. \tag{3.2}$$

Parseval's identity allows us to conclude that

$$\|p(a,t)\|^2_{L^2(\Omega)} = \sum_{j=0}^{\infty} |p^j(a,t)|^2 \quad \text{a.e. in } (0,a_\dagger) \times (0,+\infty),$$

and then we infer that

$$\|p(t)\|^2_{L^2((0,a_\dagger)\times\Omega)} = \sum_{j=0}^{\infty} \|p^j(t)\|^2_{L^2(0,a_\dagger)}$$

$$\leq \sum_{j=0}^{N} \|p^j(t)\|^2_{L^2(0,a_\dagger)} + C \sum_{j=N+1}^{\infty} \|p_0^j\|^2_{L^2(0,a_\dagger)}$$

($p_0 = \sum_{j=0}^{\infty} p_0^j \varphi_j$ in $L^2(\Omega)$ and C is a positive constant).

Since $\sum_{j=0}^{\infty} \|p_0^j\|^2_{L^2(0,a_\dagger)}$ is convergent, we conclude that for any $\varepsilon > 0$, there exists $N_\varepsilon \in \mathbf{N}$ such that

$$C \sum_{j=N_\varepsilon+1}^{\infty} \|p_0^j\|^2_{L^2(0,a_\dagger)} < \varepsilon.$$

The last relation implies

$$\|p(t)\|^2_{L^2((0,a_\dagger)\times\Omega)} \leq \varepsilon + \sum_{j=0}^{N_\varepsilon} \|p^j(t)\|^2_{L^2(0,a_\dagger)}$$

and using (3.2), we infer that

$$\limsup_{t\to+\infty} \|p(t)\|^2_{L^2((0,a_\dagger)\times\Omega)} \le \varepsilon, \quad \forall\varepsilon > 0$$

and so,

$$\lim_{t\to+\infty} \|p(t)\|^2_{L^2((0,a_\dagger)\times\Omega)} = 0.$$

THEOREM 4.3.2. *If $R = 1$ and if p_0^0 is a nontrivial datum (here $p_0^0 = (p_0, \varphi_0)$, where $(\cdot,\cdot)$ is the scalar product in $L^2(\Omega)$), then there exists*

$$\lim_{t\to+\infty} p(t) = \widetilde{p} \quad in\ L^2((0,a_\dagger)\times\Omega),$$

and

$$\widetilde{p}(a,x) = ue^{-\int_0^a \mu(s)ds}, \quad \forall(a,x)\in[0,a_\dagger]\times\overline{\Omega}, \tag{3.3}$$

$u \in (0,+\infty)$.

All the functions given by (3.3) are steady-states for (3.1).

Proof. The results in Section 2.3 allow us to conclude that there exists

$$\lim_{t\to+\infty} p^0(t) = \widetilde{p}_0 \quad \text{in } L^2(0,a_\dagger) \tag{3.4}$$

(in fact this limit exists even in $L^\infty(0,a_\dagger)$), where

$$\widetilde{p}_0(a) = u_0 \exp\{-\int_0^a \mu(s)ds\}, \quad \forall a\in(0,a_\dagger),$$

for a certain $u_0 \in (0,+\infty)$.

On the other hand

$$R_j = \int_0^{a_\dagger} \beta(a)e^{-\int_0^a (\mu(s)+\lambda_j k)ds}da < 1, \quad \forall j\in\mathbf{N}^*$$

and this implies that

$$\lim_{t\to+\infty} \|p^j(t)\|^2_{L^2(0,a_\dagger)} = 0, \quad \forall j\in N^*. \tag{3.5}$$

This yields

$$\|p(t)-\widetilde{p}\|^2_{L^2((0,a_\dagger)\times\Omega)} = \|p^0(t)-\widetilde{p}_0\|^2_{L^2(0,a_\dagger)} + \sum_{j=1}^{\infty} \|p^j(t)\|^2_{L^2(0,a_\dagger)},$$

where $\widetilde{p}(a,x) = \widetilde{p}_0(a)\varphi_0(x)$ a.e. $(a,x)\in(0,a_\dagger)\times\Omega$, and by (3.4) and (3.5) we obtain in the same manner as in the proof of Theorem 4.3.1 that

$$\lim_{t\to+\infty} p(t) = \widetilde{p} \quad \text{in } L^2((0,a_\dagger)\times\Omega)$$

and that $\widetilde{p}$ has the form in (3.3).

Finally, we establish the asymptotic behaviour in the case $R > 1$:

THEOREM 4.3.3. *If $R > 1$ and if p_0^0 is a nontrivial datum, then*

$$\lim_{t\to+\infty} \|p(t)\|_{L^2((0,a_\dagger)\times\Omega)} = +\infty. \tag{3.6}$$

Proof. By Remark 2.3.2 we may conclude that

$$\lim_{t\to\infty} \|p^0(t)\|^2_{L^2(0,a_\dagger)} = +\infty.$$

On the other hand

$$\|p(t)\|^2_{L^2((0,a_\dagger)\times\Omega)} = \sum_{j=0}^{\infty} \|p^j(t)\|^2_{L^2(0,a_\dagger)} \geq \|p^0(t)\|^2_{L^2(0,a_\dagger)},$$

$\forall t \in \mathbf{R}^+$, $t > a_\dagger$ and we conclude that (3.6) holds.

REMARK 4.3.1. *If $R \geq 1$ and if p_0^0 is a trivial datum, then the large time behaviour of p is reduced to the large time behaviour of p^1, and so on.*

REMARK 4.3.2. *Since $\varphi_0 = \dfrac{1}{meas(\Omega)^{\frac{1}{2}}}$, we conclude that if $p_0(a,x) > 0$ a.e. in $(0,a_\dagger)\times\Omega$, then p_0^0 is a nontrivial datum, and so Theorems 4.3.1, 4.3.2, 4.3.3 clarify the asymptotic behaviour of p.*

We leave to the reader the study of the asymptotic behaviour of the solution to the "separable" nonlinear model (2.2) (in Section 4.2).

Hint. Use the asymptotic behaviour of the solution to the linear problem without diffusion.

*

* *

The study of age-dependent population dynamics with diffusion has been stimulated by the work of M.E. Gurtin and R.C. MacCamy [40], [42], [44], [45] and by the paper of D.G. Aronson [13]. Results for this kind of model have been obtained by several authors. We mention in this regard the papers of S. Aniţa [6], S. Busenberg and M. Iannelli [26], M. Langlais [59], A. Leung [65].

For the semigroup approach to linear or nonlinear models with diffusion we refer to Aniţa [6], [7], Barbu and Iannelli [18], [20], Kubo [56], Webb [85], [87].

The asymptotic behaviour of the solution has been investigated for a very general nonlinear model with diffusion by M. Langlais [60]. For related results we refer to Marcati [69]. M. Kubo and M. Langlais [57] stated important results for periodic solutions.

Section 4.1. treats the basic properties of the solution to the linear model with diffusion.

The approach in Section 4.2 is based on [2] and concerns a quite general model with diffusion.

The study of the asymptotic behaviour of the solution for the linear model is inspired by Langlais [60].

4.4. Exercises

1. Solve the problem

$$\begin{aligned} &Dp + p - p_{xx} = \sin x, && (a,t,x) \in (0,1)\times(0,1)\times(0,\pi),\\ &p(a,t,0) = p(a,t,\pi) = 0, && (a,t) \in (0,1)\times(0,1),\\ &p(0,t,x) = \int_0^1 p(a,t,x)da, && (t,x) \in (0,1)\times(0,\pi),\\ &p(a,0,x) = (1-a)\sin x, && (a,x) \in (0,1)\times(0,\pi). \end{aligned}$$

Hint. The system of functions $\{\sqrt{\frac{2}{\pi}}\sin nx\}_{n\in\mathbf{N}^*}$ is a complete orthonormal system of eigenfunctions for the operator $-\frac{d^2}{dx^2}$ with homogeneous Dirichlet boundary conditions on $(0,\pi)$ and $\lambda_n = n^2$, $n \in \mathbf{N}^*$ are the corresponding eigenvalues.

The solution p is given by

$$p(a,t,x) = \sum_{n=1}^{\infty} p_n(a,t)\sqrt{\frac{2}{\pi}}\sin nx, \quad (a,t,x) \in [0,1]\times[0,1]\times[0,\pi],$$

where

$$Dp_n + (1+n^2)p_n = \begin{cases} \sqrt{\frac{\pi}{2}}, & n = 1,\\ 0, & n \geq 2, \end{cases}$$

$(a,t) \in [0,1]\times[0,1]$,

$$p_n(0,t) = \int_0^1 p_n(a,t)da \qquad t \in [0,1],$$

$$p_n(a,0) = \begin{cases} (1-a)\sqrt{\frac{\pi}{2}}, & n=1, \\ 0, & n \geq 2, \end{cases}$$

$a \in [0,1]$.

2. Solve the problem

$$Dp + \mu p - p_{xx} = \delta_1, \qquad \text{in } (0,a_\dagger) \times (0,T) \times (0,2),$$

$$p(a,t,0) = p(a,t,2) = 0, \qquad \text{in } (0,a_\dagger) \times (0,T),$$

$$p(0,t,x) = \int_0^{a_\dagger} \beta p(a,t,x)da, \quad \text{in } (0,T) \times (0,2),$$

$$p(a,0,x) = 1, \qquad \text{in } (0,a_\dagger) \times (0,2),$$

where β, $\mu > 0$ are constants and δ_1 is the Dirac distribution with support in $x := 1$.

Hint. We look for p of the form

$$p(a,t,x) = \sum_{n=1}^{\infty} p_n(a,t) \sin\frac{n\pi x}{2}, \quad (a,t) \in [0,a_\dagger] \times [0,T], \ x \in [0,2],$$

where

$$\begin{cases} Dp_n + (\mu + \frac{n^2\pi^2}{4})p_n = \sin\frac{n\pi}{2}, & (a,t) \in (0,a_\dagger) \times (0,T), \\ p_n(0,t) = \displaystyle\int_0^{a_\dagger} \beta p_n(a,t)da, & t \in (0,T), \\ p_n(a,0) = \displaystyle\int_0^2 \sin\frac{n\pi x}{2} dx, & a \in (0,a_\dagger). \end{cases}$$

3. Show that under the assumptions **(A'1)**, **(A'2)**, **(A'4)** in Section 2.1 and if in addition $f \in L^\infty((0,a_\dagger) \times \mathbf{R} \times \Omega)$, $f(a,t,x) = f(a,t+T,x)$ a.e. $(a,t,x) \in (0,a_\dagger) \times \mathbf{R} \times \Omega$ and $R = \int_0^{a_\dagger} \beta(a) e^{-\int_0^a \mu(s)ds} da < 1$, then there is a unique solution to

$$Dp(a,t,x) + \mu(a)p(a,t,x) - k\Delta p(a,t,x) = f(a,t,x), \quad \text{in } Q,$$

$$\frac{\partial p}{\partial \nu}(a,t,x) = 0, \qquad \text{on } \Sigma,$$

$$p(0,t,x) = \int_0^{a_\dagger} \beta(a)p(a,t,x)da, \qquad \text{in } \mathbf{R} \times \Omega,$$

$$p(a,t,x) = p(a,t+T,x), \qquad \text{in } Q.$$

Here $Q = (0, a_\dagger) \times \mathbf{R} \times \Omega$ and $\Sigma = (0, a_\dagger) \times \mathbf{R} \times \partial\Omega$.

Hint. Consider the orthonormal basis in $L^2(\Omega)$: $\{\varphi_j\}_{j\in\mathbf{N}} \in L^2(\Omega)$ (and $\{\lambda_j\} \subset \mathbf{R}^+$, $\lambda_0 = 0$, $\lambda_j \nearrow +\infty$ as $j \to \infty$), where

$$\begin{cases} -\Delta\varphi_j(x) = \lambda_j\varphi_j(x) & \text{in } \Omega, \\ \dfrac{\partial\varphi_j}{\partial\nu}(x) = 0 & \text{on } \partial\Omega. \end{cases}$$

We look for p as the sum of the Fourier series

$$p(a,t,x) = \sum_{j=0}^{\infty} p^j(a,t)\varphi_j(x), \quad (a,t,x) \in \overline{Q}.$$

4. Investigate the large-time behaviour of the solution to the following problem:

$$\begin{aligned} &Dp + p + \mathcal{M}\left(\int_0^1 p(a,t,x)da\right)p - \Delta p = 0, && \text{in } Q, \\ &\frac{\partial p}{\partial\nu}(a,t,x) = 0, && \text{on } \Sigma, \\ &p(0,t,x) = \int_0^1 p(a,t,x)da, && \text{in } (0,+\infty)\times\Omega, \\ &p(a,0,x) = p_0(a,x), && \text{in } (0,1)\times\Omega, \end{aligned} \tag{4.1}$$

where Ω and p_0 satisfy the hypotheses in Section 4.3 (corresponding to $a_\dagger := 1$), $Q = (0,1) \times (0,+\infty) \times \Omega$, $\Sigma = (0,1) \times (0,+\infty) \times \partial\Omega$ and $\mathcal{M} : \mathbf{R}^+ \to \mathbf{R}^+$ is a continuous and nondecreasing function.

Hint. The solution p to (4.1) satisfies

$$0 \le p(a,t,x) \le h(a,t,x) \quad \text{a.e. in } Q,$$

where h is the solution to (4.1) corresponding to $\mathcal{M} \equiv 0$.

Since $R = \displaystyle\int_0^1 e^{-a}da < 1$, it follows as in Section 4.3 that

$$\lim_{t\to+\infty} \|h(t)\|_{L^2((0,1)\times\Omega)} = 0$$

and consequently

$$\lim_{t\to+\infty} \|p(t)\|_{L^2((0,1)\times\Omega)} = 0.$$

5. Solve the problem

$$\begin{aligned}
&Dp + \frac{1}{1-a}p - p_{xx} = 0, && \text{in } Q,\\
&p_x(a,t,0) = p_x(a,t,2\pi) = 0, && \text{on } (0,1)\times(0,+\infty),\\
&p(0,t,x) = 2\int_0^1 p(a,t,x)da, && \text{in } \mathbf{R}\times(0,2\pi),\\
&p(a,0,x) = (1-a)\cos x, && \text{in } (0,1)\times(0,2\pi).
\end{aligned}$$

Here $Q = (0,1)\times(0,+\infty)\times(0,2\pi)$.

Hint. The solution p satisfies

$$p(a,t,x) = \tilde{p}(a,t)h(t,x) \qquad \text{a.e.,}$$

where $\tilde{p}$ is the solution to

$$\begin{aligned}
&Dp + \frac{1}{1-a}p = 0, && (a,t)\in(0,1)\times(0,+\infty),\\
&p(0,t) = 2\int_0^1 p(a,t)da, && t\in(0,+\infty),\\
&p(a,0) = 1-a, && a\in(0,1),
\end{aligned}$$

and h is the solution to

$$\begin{aligned}
&h_t(t,x) - h_{xx}(t,x) = 0, && (t,x)\in(0,+\infty)\times(0,2\pi),\\
&h_x(t,0) = h_x(t,2\pi) = 0, && t\in(0,+\infty),\\
&h(0,x) = \cos x, && x\in(0,2\pi).
\end{aligned}$$

6. Assume that Ω, k, $a_\dagger$, β, μ and p_0 satisfy **(Ai1)-(Ai3)** in Section 4.1. Investigate the large-time behaviour of the solution to the following problem

$$\begin{aligned}
&Dp(a,t,x) + \mu(a)p(a,t,x) - k\Delta p(a,t,x) = 0, && \text{in } Q,\\
&p(a,t,x) = 0, && \text{on } \Sigma,\\
&p(0,t,x) = \int_0^{a_\dagger} \beta(a)p(a,t,x)da, && \text{in } (0,+\infty)\times\Omega,\\
&p(a,0,x) = p_0(a,x), && \text{in } (0,a_\dagger)\times\Omega.
\end{aligned}$$

Here $Q = (0, a_\dagger) \times \mathbf{R} \times \Omega$ and $\Sigma = (0, a_\dagger) \times \mathbf{R} \times \partial\Omega$.

Hint. Use the same argument as in Section 4.3.

CHAPTER 5

CONTROL OF POPULATION DYNAMICS WITH DIFFUSION

This chapter is devoted to the optimal harvesting problem governed by nonlinear age-dependent population dynamics with diffusion and to some controllability problems with distributed parameter. The main goal of the first section is to establish the existence of an optimal control and to obtain first order necessary conditions of optimality for the optimal harvesting problem. These conditions allow calculation or approximation of the optimal control and the optimal harvest. The next two sections concern controllability for the linear model with controllers acting in the whole domain, and for this case we shall indicate a feedback control, or acting in a subdomain, and in this case we will establish the existence of a control which realizes the goal.

The models to be studied describe a biological population which is free to move in a habitat $\Omega \subset \mathbf{R}^n$ ($n \in \{1,2,3\}$). The habitat is a nonempty bounded domain with a smooth enough boundary $\partial\Omega$. The study concerns models with homogeneous Neumann boundary conditions. By similar methods we can treat the models with homogeneous Dirichlet boundary conditions.

Throughout this chapter $a_\dagger$, T, $k \in (0,+\infty)$, $Q_T = (0,a_\dagger) \times (0,T) \times \Omega$, $\Sigma_T = (0,a_\dagger) \times (0,T) \times \partial\Omega$ and Dp is the directional derivative

$$Dp(a,t,x) = \lim_{\varepsilon\to 0} \frac{p(a+\varepsilon,t+\varepsilon,x) - p(a,t,x)}{\varepsilon}.$$

5.1. Optimal harvesting for a general nonlinear model

This section concerns a general optimal harvesting problem in the time interval $(0,T)$, $T > 0$, subject to an external supply of individuals $f(a,t,x) \geq 0$ and to a specific harvesting effort $u(a,t,x)$.

Consider the following system

$$\begin{cases} Dp + \mu(a,t,x,P(t,x))p - k\Delta p = f - up, & \text{in } Q_T, \\ \dfrac{\partial p}{\partial \nu}(a,t,x) = 0, & \text{on } \Sigma_T, \\ p(0,t,x) = \displaystyle\int_0^{a_\dagger} \beta(a,t,x,P(t,x))p(a,t,x)\,da, & \text{in } (0,T)\times\Omega, \\ p(a,0,x) = p_0(a,x), & \text{in } (0,a_\dagger)\times\Omega. \end{cases} \tag{1.1}$$

We deal with the problem of finding the harvesting effort u in order to obtain the best harvest (the best cost of the harvest):

$$\textbf{(OH)} \qquad Maximize \int_{Q_T} u(a,t,x)g(a,t,x)p^u(a,t,x)dx\ dt\ da,$$

subject to $u \in \mathcal{U}$,

$$\mathcal{U} = \{v \in L^2(Q_T);\ \zeta_1(a,t,x) \le v(a,t,x) \le \zeta_2(a,t,x)\ a.e.\ (a,t,x) \in Q_T\}$$

(here ζ_1, $\zeta_2 \in L^\infty(Q_T)$, $0 \le \zeta_1(a,t,x) \le \zeta_2(a,t,x)$ a.e. in Q_T), where p^u is the solution of (1.1).

From a biological point of view $g(a,t,x) \ge 0$ is a weight (the price of an individual of age a at the moment t and location x).

One assumes the hypotheses **(An1)–(An4)** in Section 4.2 hold and that

$$\textbf{(An5)} \qquad g \in L^\infty(Q_T),\ g(a,t,x) \ge 0 \quad \text{a.e. } (a,t,x) \in Q_T.$$

By Theorem 4.2.1 it follows that for any $u \in \mathcal{U}$ there exists a unique solution p^u to (1.1). The solution p^u belongs to $L^\infty(Q_T)$.

A COMPACTNESS RESULT FOR (1.1)

For any $u \in \mathcal{U}$ we define

$$P^u(t,x) = \int_0^{a_\dagger} p^u(a,t,x)da, \qquad (t,x) \in (0,T)\times\Omega.$$

LEMMA 5.1.1. *The set* $\{P^u;\ u \in \mathcal{U}\}$ *is relatively compact in* $L^2((0,T)\times\Omega)$.

Proof. Since p^u is a solution of (1.1), then for any $\varepsilon > 0$ small enough we have that $P^{u,\varepsilon}$, defined by

$$P^{u,\varepsilon}(t,x) = \int_0^{a_\dagger - \varepsilon} p^u(a,t,x)da, \quad (t,x) \in (0,T) \times \Omega,$$

is a solution to

$$P_t^{u,\varepsilon} - k\Delta P^{u,\varepsilon} = \int_0^{a_\dagger - \varepsilon} [f - (\mu(a,t,x,P^u(t,x)) + u)\, p^u(a,t,x)]\, da$$

$$-p^u(a_\dagger - \varepsilon, t, x) + \int_0^{a_\dagger} \beta(a,t,x,P^u(t,x))p^u \, da, \ \text{ in } (0,T) \times \Omega,$$

$$\frac{\partial P^{u,\varepsilon}}{\partial \nu}(t,x) = 0, \qquad \text{on } (0,T) \times \partial\Omega,$$

$$P^{u,\varepsilon}(0,x) = \int_0^{a_\dagger - \varepsilon} p_0(a,x)da, \ \text{ in } \Omega\, .$$

Since $\{up^u\}$, $\{\mu(\cdot,\cdot,\cdot,P^u)p^u\}$ are bounded in $L^\infty((0,a_\dagger - \varepsilon) \times (0,T) \times \Omega)$, $\{\beta(\cdot,\cdot,\cdot,P^u)p^u\}$ is bounded in $L^\infty(Q_T)$ and $\{p^u(a_\dagger - \varepsilon,\cdot,\cdot)\}$ is bounded in $L^\infty((0,T) \times \Omega)$ - with respect to $u \in \mathcal{U}$ (as a consequence of Theorem 4.2.2), we conclude that $\{P_t^{u,\varepsilon} - k\Delta P^{u,\varepsilon}\}$ is bounded in $L^\infty((0,T) \times \Omega)$ and in conclusion $\{P^{u,\varepsilon};\ u \in \mathcal{U}\}$ is bounded in $L^2(0,T;H_0^1(\Omega))$ and $\{P_t^{u,\varepsilon};\ u \in \mathcal{U}\}$ is bounded in $L^2(0,T;H^{-1}(\Omega))$.

The following compactness result is useful in our context (see [15], [84]):

THEOREM (J.P. Aubin). *Let X_0, X_1 and X_2 be three Banach spaces such that*

$$X_0 \subset X_1 \subset X_2$$

(algebraically and topologically), X_0, X_1 are reflexives and the injection of X_0 into X_1 is compact. Let r_0, $r_1 \in (1,+\infty)$ and a, $b \in \mathbf{R}$, $a < b$. Then the space

$$L^{r_0}(a,b;X_0) \cap W^{1,r_1}(a,b;X_2)$$

is compactly imbedded in $L^{r_0}(a,b;X_1)$.

By this theorem we get that for any $\varepsilon > 0$ small enough, the set $\{P^{u,\varepsilon};\ u \in \mathcal{U}\}$ is relatively compact in $L^2((0,T) \times \Omega)$.

On the other hand

$$|P^{u,\varepsilon}(t,x) - P^u(t,x)| \le \int_{a_\dagger - \varepsilon}^{a_\dagger} |p^u(a,t,x)|da \le \varepsilon \|\overline{p}\|_{L^\infty(Q_T)},$$

$\forall \varepsilon > 0$, $\forall u \in \mathcal{U}$, a.e. $(t,x) \in (0,T) \times \Omega$, where $\overline{p}$ is the solution of (1.1) corresponding to $\mu := 0$, $u := 0$, $\beta := \|\beta(\cdot,\cdot,\cdot,0)\|_{L^\infty(Q_T)}$, $p_0 := \|p_0\|_{L^\infty((0,a_\dagger)\times\Omega)}$.

Combining these we may conclude the relative compactness of $\{P^u;\ u \in \mathcal{U}\}$ in $L^2(Q_T)$.

EXISTENCE OF AN OPTIMAL CONTROL

THEOREM 5.1.2. *Problem* **(OH)** *admits at least one optimal pair.*

Proof. Let $\Psi : \mathcal{U} \to \mathbf{R}^+$ be defined by

$$\Psi(u) = \int_{Q_T} u(a,t,x) g(a,t,x) p^u(a,t,x) dx\ dt\ da$$

and let

$$d = \sup_{u \in \mathcal{U}} \Psi(u)\ .$$

Since by Theorem 4.2.2

$$0 \le \Psi(u) \le \int_{Q_T} \zeta_2(a,t,x) g(a,t,x) \overline{p}(a,t,x)\ dx\ dt\ da$$

(with $\overline{p}$ defined above), we have $d \in [0,+\infty)$. Now let $\{u_N\} \subset \mathcal{U}$, $N \in \mathbf{N}^*$, be a sequence such that

$$d - \frac{1}{N} < \Psi(u_N) \le d\ .$$

Since

$$0 \le p^{u_N}(a,t,x) \le \overline{p}(a,t,x) \quad \text{a.e. in } Q_T\ ,$$

we conclude that there exists a subsequence, also denoted by $\{u_N\}_{N \in \mathbf{N}^*}$, such that

$$p^{u_N} \to p^* \quad \text{weakly in } L^2(Q_T).$$

Lemma 5.1.1 implies the existence of a subsequence, also denoted by $\{P^{u_N}\}_{N \in N^*}$ such that

$$\begin{aligned} &P^{u_N} \to P^* && \text{in } L^2((0,T) \times \Omega), \\ &P^{u_N}(t,x) \to P^*(t,x) && \text{a.e. in } (0,T) \times \Omega. \end{aligned} \tag{1.2}$$

Using now Mazur's theorem we obtain the existence of a sequence $\{\widetilde{p}_N\}_{N \in N^*}$ such that

$$\widetilde{p}_N(a,t,x) = \sum_{i=N+1}^{k_N} \lambda_i^N p^{u_i}(a,t,x), \quad \lambda_i^N \ge 0, \quad \sum_{i=N+1}^{k_N} \lambda_i^N = 1$$

$(k_N \geq N+1)$ and

$$\widetilde{p}_N \to p^* \quad \text{in } L^2(Q_T)$$

as $N \to +\infty$.

Consider now the sequence of controls

$$\widetilde{u}_N(a,t,x) = \begin{cases} \dfrac{\sum_{i=N+1}^{k_N} \lambda_i^N u_i(a,t,x) p^{u_i}(a,t,x)}{\sum_{i=N+1}^{k_N} \lambda_i^N p^{u_i}(a,t,x)}, & \text{if } \displaystyle\sum_{i=N+1}^{k_N} \lambda_i^N p^{u_i}(a,t,x) \neq 0, \\ \zeta_1(a,t,x), & \text{if } \displaystyle\sum_{i=N+1}^{k_N} \lambda_i^N p^{u_i}(a,t,x) = 0\ . \end{cases}$$

For these controls we have

$$\widetilde{u}_N \in \mathcal{U}\ .$$

Since $p^{u_N} \to p^*$ weakly in $L^2(Q_T)$, we obtain that

$$\int_0^{a_\dagger} p^{u_N}(a,\cdot,\cdot)da \to \int_0^{a_\dagger} p^*(a,\cdot,\cdot)da \quad \text{weakly in } L^2((0,T)\times\Omega)$$

and consequently

$$P^*(t,x) = \int_0^{a_\dagger} p^*(a,t,x)da \quad \text{a.e. in } (0,T)\times\Omega.$$

There exists a subsequence, also denoted by $\{\widetilde{u}_N\}_{N\in N^*}$, such that

$$\widetilde{u}_N \to u^* \quad \text{weakly in } L^2(Q_T)$$

$(u^* \in \mathcal{U})$. It is obvious now that $\widetilde{p}_N$ is a solution of

$$\begin{cases} Dp - k\Delta p + \displaystyle\sum_{i=N+1}^{k_N} \lambda_i^N \mu(a,t,x,P^{u_i}(t,x))p^{u_i} & \\ \qquad = f - \widetilde{u}_N p, & \text{in } Q_T, \\ \dfrac{\partial p}{\partial \nu}(a,t,x) = 0, & \text{on } \Sigma_T, \\ p(0,t,x) = \displaystyle\int_0^{a_\dagger} \sum_{i=N+1}^{k_N} \lambda_i^N \beta(a,t,x,P^{u_i}(t,x))p^{u_i} & da, \\ & \text{in } (0,T)\times\Omega, \\ p(a,0,x) = p_0(a,x), & \text{in } (0,a_\dagger)\times\Omega. \end{cases} \tag{1.3}$$

By (1.2) we deduce the existence of a subsequence (also denoted by $\{u_N\}$) such that

$$\begin{cases} \mu(a,t,x,P^{u_N}(t,x)) \to \mu(a,t,x,P^*(t,x) \text{ a.e. in } Q_T, \\ \beta(a,t,x,P^{u_N}(t,x)) \to \beta(a,t,x,P^*(t,x)) \text{ a.e. in } Q_T. \end{cases} \tag{1.4}$$

Thus by (1.4) and since $\tilde{p}_N \to p^*$ in $L^2(Q_T)$ we have that

$$\sum_{i=N+1}^{k_N} \lambda_i^N \mu(a,t,x,P^{u_i}(t,x))p^{u_i}(a,t,x) \to \mu(a,t,x,P^*(t,x))p^*(a,t,x)$$

a.e. in Q_T and

$$\sum_{i=N+1}^{k_N} \lambda_i^N \beta(a,t,x,P^{u_i}(t,x))p^{u_i}(a,t,x) \to \beta(a,t,x,P^*(t,x))p^*(a,t,x)$$

a.e. in Q_T.

Passing to the limit in (1.3) (on the characteristic lines) we obtain that p^* is the solution of (1.1) corresponding to u^*.

Moreover we have

$$\sum_{i=N+1}^{k_N} \lambda_i^N \int_{Q_T} u_i(a,t,x)g(a,t,x)p^{u_i}(a,t,x)dx\,dt\,da$$

$$= \int_{Q_T} \widetilde{u}_N(a,t,x)g(a,t,x)\widetilde{p}_N(a,t,x)dx\,dt\,da$$

$$= \sum_{i=N+1}^{k_N} \lambda_i^N \Psi(u_i) \to \Psi(u^*)$$

(as $N \to +\infty$). We may infer now that $d = \Psi(u^*)$.

NECESSARY OPTIMALITY CONDITIONS

Assume in addition that the derivative of β and μ with respect to P (β_P, μ_P) exist and belong to $L^\infty(Q_T \times [0,+\infty))$.

THEOREM 5.1.3. If (u^*, p^{u^*}) is an optimal pair for **(OH)** and if q is the solution of (1.5)-(1.5)'

$$Dq(a,t,x) + k\Delta q(a,t,x) - \mu(a,t,x,P^{u^*}(t,x))q(a,t,x)$$

$$- \int_0^{a_\dagger} \mu_P(a,t,x,P^{u^*}(t,x))p^{u^*}(a,t,x)q(a,t,x)da$$

$$+\left(\beta(a,t,x,P^{u^*}(t,x)) + \int_0^{a_\dagger} \beta_P(a,t,x,P^{u^*}(t,x))p^{u^*}(a,t,x)da\right) q(0,t,x)$$

$$= u^*(a,t,x)(g+q)(a,t,x), \qquad (a,t,x) \in Q_T, \tag{1.5}$$

$$\begin{cases} \dfrac{\partial q}{\partial \nu}(a,t,x) = 0, & (a,t,x) \in \Sigma_T, \\ q(a_\dagger,t,x) = 0, & (t,x) \in (0,T) \times \Omega, \\ q(a,T,x) = 0, & (a,x) \in (0,a_\dagger) \times \Omega, \end{cases} \qquad (1.5)$$

then we have

$$u^*(a,t,x) = \begin{cases} \zeta_1(a,t,x) & \text{if } (g+q)(a,t,x) < 0, \\ \zeta_2(a,t,x) & \text{if } (g+q)(a,t,x) > 0. \end{cases}$$

Proof. Existence and uniqueness of q, a solution of (1.5)-(1.5)$'$ follows in the same way as the existence and uniqueness of the solution of (1.1). Since (u^*, p^{u^*}) is an optimal pair for **(OH)** we get

$$\int_{Q_T} u^*(a,t,x)g(a,t,x)p^{u^*}(a,t,x)dx\ dt\ da$$

$$\geq \int_{Q_T} (u^*(a,t,x) + \delta v(a,t,x))g(a,t,x)p^{u^*+\delta v}(a,t,x)dx\ dt\ da\ ,$$

for any $\delta > 0$ small enough, $\forall v \in L^\infty(Q_T)$ such that

$$\begin{cases} v(a,t,x) \leq 0 & \text{if } u^*(a,t,x) = \zeta_2(a,t,x), \\ v(a,t,x) \geq 0 & \text{if } u^*(a,t,x) = \zeta_1(a,t,x), \end{cases}$$

which implies

$$\int_{Q_T} u^*(a,t,x)g(a,t,x)\frac{p^{u^*+\delta v}(a,t,x) - p^{u^*}(a,t,x)}{\delta}dx\ dt\ da$$

$$+\int_{Q_T} v(a,t,x)g(a,t,x)p^{u^*+\delta v}(a,t,x)dx\ dt\ da \leq 0\ . \tag{1.6}$$

Using the definition of the solution to (1.1) and the comparision result in Section 4.1 we may infer that for any $v \in L^\infty(Q_T)$ as above, the following convergence holds

$$p^{u^*+\delta v} \longrightarrow p^{u^*} \text{ in } L^\infty(0,T;L^2((0,a_\dagger)\times\Omega)), \tag{1.7}$$

as $\delta \longrightarrow 0^+$. Indeed, define

$$z^\delta(a,t,x) = \frac{p^{u^*+\delta v}(a,t,x) - p^{u^*}(a,t,x)}{\delta}, \quad (a,t,x) \in Q_T$$

and

$$y^\delta = \delta z^\delta.$$

The function y^δ is a solution of

$$\begin{cases} Dy^\delta - k\Delta y^\delta + \mu(a,t,x,P^{u^*+\delta v}(t,x))p^{u^*+\delta v} - \mu(a,t,x,P^{u^*}(t,x))p^{u^*} \\ \quad = -u^* y^\delta - \delta v(a,t,x)p^{u^*+\delta v}, \qquad (a,t,x) \in Q_T, \\ \dfrac{\partial y^\delta}{\partial \nu}(a,t,x) = 0, \qquad (a,t,x) \in \Sigma_T, \\ y^\delta(0,t,x) = \displaystyle\int_0^{a_\dagger} (\beta(a,t,x,P^{u^*+\delta v}(t,x))p^{u^*+\delta v} \\ \qquad -\beta(a,t,x,P^{u^*}(t,x))p^{u^*})da, \quad (t,x) \in (0,T) \times \Omega, \\ y^\delta(a,0,x) = 0, \qquad (a,x) \in (0,a_\dagger) \times \Omega. \end{cases}$$

Multiplying the first equation by y^δ and integrating on $(0,a_\dagger) \times (0,t) \times \Omega$ we obtain

$$\|y^\delta(t)\|^2_{L^2((0,a_\dagger)\times\Omega)} \le C \int_0^t \|y^\delta(s)\|^2_{L^2((0,a_\dagger)\times\Omega)} ds$$

$$+\delta \int_0^t \int_0^{a_\dagger} \int_\Omega \zeta_2(a,s,x)\overline{p}(a,s,x)|y^\delta(a,s,x)|dx\ da\ ds,$$

where C is a positive constant depending on the Lipschitz constant L(M) (where $M = Ess\ sup \displaystyle\int_0^{a_\dagger} \overline{p}(a,t,x)da$, $(t,x) \in (0,T) \times \Omega$). This implies that

$$\|y^\delta(t)\|^2_{L^2((0,a_\dagger)\times\Omega)} \le (1+C) \int_0^t \|y^\delta(t)\|^2_{L^2((0,a_\dagger)\times\Omega)}$$

$$+\delta^2 \int_{Q_T} \zeta_2^2(a,t,x)\overline{p}^2(a,t,x)dx\ da\ dt.$$

By Bellman's lemma we deduce that

$$\|y^\delta(t)\|^2_{L^2((0,a_\dagger)\times\Omega)} \le \delta^2 e^{(1+C)t} \int_{Q_T} \zeta_2^2(a,t,x)\overline{p}^2(a,t,x)dx\ da\ dt,$$

for any $t \in [0,T]$. Passing to the limit in the last inequality we get

$$y^\delta \to 0 \quad \text{in } L^\infty(0,T; L^2((0,a_\dagger) \times \Omega))$$

as $\delta \to 0^+$ and so (1.7) follows.

In the same manner as above (and using (1.7)) it follows that

$$z^\delta \longrightarrow z$$

in $L^2(Q_T)$ as $\delta \longrightarrow 0$, where z is the solution of

$$\begin{cases} Dz - k\Delta z + \mu(a,t,x,P^{u^*}(t,x))z(a,t,x) \\ +\mu_P(a,t,x,P^{u^*}(t,x))p^{u^*}(a,t,x)\int_0^{a_\dagger} z(a,t,x)da \\ = -u^*z - v(a,t,x)p^{u^*}, & \text{in } Q_T, \\ \frac{\partial z}{\partial \nu}(a,t,x) = 0, & \text{on } \Sigma_T, \\ z(0,t,x) = \int_0^{a_\dagger} \beta(a,t,x,P^{u^*}(t,x))z(a,t,x)da \\ \quad & \text{in } (0,T)\times\Omega, \\ +\int_0^{a_\dagger}\left(\beta_P(a,t,x,P^{u^*}(t,x))p^{u^*}\int_0^{a_\dagger} z(s,t,x)ds\right)da, \\ z(a,0,x) = 0, & \text{in } (0,a_\dagger)\times\Omega. \end{cases}$$

Passing to the limit in (1.6) ($\delta \to 0^+$) we conclude that

$$\int_{Q_T} u^*(a,t,x)g(a,t,x)z(a,t,x)dx\ dt\ da$$
$$+\int_{Q_T} v(a,t,x)g(a,t,x)p^{u^*}(a,t,x)dx\ dt\ da \le 0,$$

$\forall v \in L^\infty(Q_T)$ such that

$$\begin{cases} v(a,t,x) \le 0 & \text{if } u^*(a,t,x) = \zeta_2(a,t,x), \\ v(a,t,x) \ge 0 & \text{if } u^*(a,t,x) = \zeta_1(a,t,x). \end{cases}$$

Multiplying (1.5) by z and integrating over Q_T we get after some calculation that

$$\int_{Q_T} (u^*gz)(a,t,x)dx\ dt\ da = \int_{Q_T} (vp^{u^*}q)(a,t,x)dx\ dt\ da,$$

and consequently

$$\int_{Q_T} v(a,t,x)p^{u^*}(a,t,x)(g+q)(a,t,x)dx\ dt\ da \le 0,$$

$\forall v \in L^{\infty}(Q_T)$ such that

$$\begin{cases} v(a,t,x) \le 0 & \text{if } u^*(a,t,x) = \zeta_2(a,t,x), \\ v(a,t,x) \ge 0 & \text{if } u^*(a,t,x) = \zeta_1(a,t,x), \end{cases}$$

which implies

$$p^{u^*}(g+q) \in N_{\mathcal{U}}(u^*)$$

(here $N_{\mathcal{U}}(u^*)$ is the normal cone at $\mathcal{U}$ in u^* (in $L^2(Q_T)$; see Appendix 1).

For any $(a,t,x) \in Q_T$ such that $p^{u^*}(a,t,x) \neq 0$, we conclude that

$$u^*(a,t,x) = \begin{cases} \zeta_1(a,t,x) & \text{if } (g+q)(a,t,x) < 0, \\ \zeta_2(a,t,x) & \text{if } (g+q)(a,t,x) > 0, \end{cases}$$

$(a,t,x) \in Q_T$. On the other hand, we define

$$B = \{(x,t,a) \in Q_T;\ p^{u^*}(a,t,x) = 0\}\ .$$

It follows that $f(a,t,x) = 0$ a.e. in B. Using now the unique continuation theorem of Mizohata (for linear parabolic equations) (see Mizohata [73]; or we can use the Carleman estimates in Section 5.3) and the reverse uniqueness theorem of Bardos and Tartar (for linear parabolic equations; see Bardos and Tartar [22]) we deduce that

$$p^{u^*}(a,t,x) = 0$$

for any $(a,t,x) \in Q_T$ for which there exists $(\tilde{a},\tilde{t},\tilde{x}) \in B$ such that $\tilde{t}-\tilde{a} = t-a$.

It is obvious now that we can change the control u^* on B with arbitrary values in $[\zeta_1(a,t,x), \zeta_2(a,t,x)]$ and the state corresponding to this new control is the same and the value of the cost functional also remains the same.

The conclusion of Theorem 5.1.3 is now obvious.

In the particular case of the vital rates independent of P, Theorem 5.1.3 implies

THEOREM 5.1.4. If (u^*, p^{u^*}) is an optimal pair for **(OH)** and if q is the solution of

$$\begin{cases} Dq + k\Delta q - \mu(a,t,x)q + \beta(a,t,x) \quad q(0,t,x) \\ \quad = u^*(g+q)(a,t,x), & (a,t,x) \in Q_T, \\ \dfrac{\partial q}{\partial \nu}(a,t,x) = 0, & (a,t,x) \in \Sigma_T, \\ q(a_\dagger, t, x) = 0, & (t,x) \in (0,T) \times \Omega, \\ q(a,T,x) = 0, & (a,x) \in (0,a_\dagger) \times \Omega, \end{cases} \tag{1.8}$$

then we have

$$u^*(a,t,x) = \begin{cases} \zeta_1(a,t,x) & \text{if } (g+q)(a,t,x) < 0, \\ \zeta_2(a,t,x) & \text{if } (g+q)(a,t,x) > 0. \end{cases}$$

As a consequence of Theorem 5.1.4 we obtain that q (the solution of (1.8)) is also a solution of

$$\begin{cases} Dq + k\Delta q - \mu q = \zeta_1(g+q) + \quad (\zeta_2 - \zeta_1)(g+q)^+ \\ \quad -\beta(a,t,x)q(0,t,x), & \text{in } Q_T, \\ \dfrac{\partial q}{\partial \nu}(a,t,x) = 0, & \text{on } \Sigma_T, \\ q(a_\dagger, t, x) = 0, & \text{in } (0,T) \times \Omega, \\ q(a,T,x) = 0, & (0,a_\dagger) \times \Omega. \end{cases} \tag{1.9}$$

In fact (1.9) has a unique solution. System (1.9) does not depend on p_0 and on f. As a consequence of this fact we obtain that q and the optimal control do not depend on p_0 and f.

REMARK 5.1.1. *If the vital rates do not depend on P, then (1.8) and (1.9) allow us to calculate an optimal effort u^*.*

If $g \equiv 1$ we leave it to the reader to find sufficient conditions in order to have a bang-bang control u^* (as in Section 3.1).

5.2. A null controllability problem with distributed parameter

This section concerns the null controllability for the linear model with diffusion. Namely, we shall prove that for any $T \in (0, +\infty)$, there exists $u \in$

$L^\infty(Q_T)$ such that the solution p^u of (2.1) to satisfy $p^u(a,T,x) = 0$ a.e. $(a,x) \in (0,a_\dagger) \times \Omega$ and $p^u(a,t,x) \geq 0$ a.e. $(a,t,x) \in Q_T$.

$$\begin{cases} Dp + \mu(a)p - k\Delta p = u(a,t,x), & (a,t,x) \in Q_T, \\ \dfrac{\partial p}{\partial \nu}(a,t,x) = 0, & (a,t,x) \in \Sigma_T, \\ p(0,t,x) = \displaystyle\int_0^{a_\dagger} \beta(a)p(a,t,x)da, & (t,x) \in (0,T) \times \Omega, \\ p(a,0,x) = p_0(a,x), & (a,x) \in (0,a_\dagger) \times \Omega. \end{cases} \tag{2.1}$$

For the sake of simplicity we assume that the vital rates depend only on age. Here are the hypotheses on β , μ and p_0:

(A'1) $\beta \in L^\infty(0,a_\dagger)$, $\beta(a) \geq 0$ a.e. $a \in (0,a_\dagger)$;

(A'2) There exists $a_0 \in (0,a_\dagger)$ such that $\beta(a) = 0$ a.e. $a \in (0,a_0)$;

(A'3) $\mu \in L^1_{loc}([0,a_\dagger))$, $\mu(a) \geq 0$ a.e. $a \in (0,a_\dagger)$;

(A'4) $\displaystyle\int_0^{a_\dagger} \mu(a)da = +\infty$;

(A'5) $p_0 \in L^\infty((0,a_\dagger) \times \Omega)$, $p_0(a,x) \geq 0$ a.e. in $(0,a_\dagger) \times \Omega$.

Denote by

$$sgn\ \gamma = \begin{cases} 1, & \text{if } \gamma > 0, \\ [-1,1], & \text{if } \gamma = 0, \\ -1, & \text{if } \gamma < 0. \end{cases}$$

We shall first study the system (2.1) with the feedback control $u := -\rho \cdot sgn\ p$, i.e.,

$$\begin{cases} Dp + \mu(a)p - k\Delta p \in -\rho \cdot sgn\ p(a,t,x), & (a,t,x) \in Q_T, \\ \dfrac{\partial p}{\partial \nu}(a,t,x) = 0, & (a,t,x) \in \Sigma_T, \\ p(0,t,x) = \displaystyle\int_0^{a_\dagger} \beta(a)p(a,t,x)da, & (t,x) \in (0,T) \times \Omega, \\ p(a,0,x) = p_0(a,x), & (a,x) \in (0,a_\dagger) \times \Omega. \end{cases} \tag{2.2}$$

Here, $\rho \in (0,+\infty)$ is a constant.

We say that p is *a solution* of (2.2) if there exists $u \in L^\infty(Q_T)$ such that $u(a,t,x) \in -\rho \cdot sgn\ p(a,t,x)$ a.e. $(a,t,x) \in Q_T$ and p is the solution of (2.1).

THEOREM 5.2.1. *There exists a unique solution p to (2.1). The solution is nonnegative and belongs to $L^\infty(Q_T)$.*

Proof. For any $\varepsilon > 0$ arbitrary but fixed we consider the following system

$$\begin{cases} Dp + \mu(a)p - k\Delta p + \rho \cdot \dfrac{p(a,t,x)}{|p(a,t,x)| + \varepsilon} = 0, & (a,t,x) \in Q_T, \\ \dfrac{\partial p}{\partial \nu}(a,t,x) = 0, & (a,t,x) \in \Sigma_T, \\ p(0,t,x) = \displaystyle\int_0^{a_\dagger} \beta(a)p(a,t,x)da, & (t,x) \in (0,T) \times \Omega, \\ p(a,0,x) = p_0(a,x), & (a,x) \in (0,a_\dagger) \times \Omega. \end{cases} \tag{2.3}$$

Since the function $f_\varepsilon : \mathbf{R} \to \mathbf{R}$, $f_\varepsilon(t) = \dfrac{t}{|t| + \varepsilon}$, is Lipschitz continous, the Banach fixed point theorem implies the existence of a unique (and nonnegative) solution p_ε to (2.3).

Since

$$-\rho \frac{p_\varepsilon(a,t,x)}{|p_\varepsilon(a,t,x)| + \varepsilon} \le 0 \quad \text{a.e. } (a,t,x) \in Q_T,$$

the comparison Theorem 4.1.4 implies that

$$0 \le p_\varepsilon(a,t,x) \le \widetilde{p}(a,t,x) \quad \text{a.e. } (a,t,x) \in Q_T, \tag{2.4}$$

where $\widetilde{p}$ is the solution of (2.1) corresponding to $u := 0$. $\widetilde{p} \in L^\infty(Q_T)$ (by Theorem 4.1.4).

Moreover, if p_{01}, p_{02} satisfy **(A'5)** and $p_{01}(a,x) \le p_{02}(a,x)$ a.e. $(a,x) \in (0,a_\dagger) \times \Omega$, then

$$p_\varepsilon(a,t,x;p_{01}) \le p_\varepsilon(a,t,x;p_{02})$$

a.e. $(a,t,x) \in Q_T$, where $p_\varepsilon(\cdot,\cdot,\cdot,p_{0i})$ is the solution of (2.3) corresponding to $p_0 := p_{0i}$, $i \in \{1,2\}$ (this follows in the same maner as Theorem 4.1.4; we use the monotonicity of f_ε).

Using again the monotonicity of f_ε and comparison Theorem 4.1.4 we get that $0 < \varepsilon_1 < \varepsilon_2$ implies

$$p_{\varepsilon_1}(a,t,x) \le p_{\varepsilon_2}(a,t,x) \quad \text{a.e. } (a,t,x) \in Q_T$$

and so

$$p_\varepsilon \to p \quad \text{in } L^2(Q_T)$$

as $\varepsilon \to 0^+$. On the other hand

$$\left\{ \frac{p_\varepsilon}{|p_\varepsilon| + \varepsilon} \right\} \quad \text{is bounded in } L^\infty(Q_T)$$

$\left(0 \leq \rho \dfrac{p_\varepsilon(a,t,x)}{|p_\varepsilon(a,t,x)|+\varepsilon} < \rho \text{ a.e. } (a,t,x) \in Q_T\right)$. We conclude that there exists a sequence $\{\varepsilon_N\}_{N\in\mathbf{N}^*}$, $\varepsilon_N \searrow 0$, such that

$$p_{\varepsilon_N} \to p \quad \text{strongly in } L^2(Q_T) \tag{2.5}$$

and

$$h_N = \frac{p_{\varepsilon_N}}{|p_{\varepsilon_N}| + \varepsilon_N} \to h \quad \text{weakly in } L^2(Q_T)$$

as $N \to +\infty$. The Mazur theorem implies that there exists

$$\widetilde{h}_N = \sum_{i=N+1}^{k_N} \lambda_i^N h_i \to h \quad \text{strongly in } L^2(Q_T),$$

where $\lambda_i^N \in [0,1]$, $\sum_{i=N+1}^{k_N} \lambda_i^N = 1$ ($\{\tilde{h}_N\}$ is a sequence of convex combinations).

Convergence (2.5) implies that

$$\widetilde{p}_N = \sum_{i=N+1}^{k_N} \lambda_i^N p_{\varepsilon_i} \to p \quad \text{strongly in } L^2(Q_T).$$

It is obvious that $\widetilde{p}_N$ is the solution of (2.1), corresponding to $u := -\rho\widetilde{h}_N$, i.e., $\widetilde{p}_N = p^{-\rho\widetilde{h}_N}$. This yields $-\rho\widetilde{h}_N \to -\rho h$ in $L^2(Q_T)$ $(N \to +\infty)$.

By Theorem 4.1.5 we may conclude that p is the solution of (2.1) corresponding to $u := -\rho h$.

Let us prove that

$$h(a,t,x) \in sgn\ p(a,t,x) \quad \text{a.e. } (a,t,x) \in Q_T.$$

Indeed, on a subsequence, also denoted by $\left\{\widetilde{h}_N\right\}_{N\in\mathbf{N}^*}$, we get the convergence

$$\widetilde{h}_N(a,t,x) \to h(a,t,x) \quad \text{a.e. } (a,t,x) \in Q_T.$$

Similarly

$$p_{\varepsilon_N}(a,t,x) \to p(a,t,x) \quad \text{a.e. } (a,t,x) \in Q_T$$

and consequently

$$\widetilde{p}_N(a,t,x) \to p(a,t,x) \quad \text{a.e. } (a,t,x) \in Q_T.$$

If, for such a $(a,t,x) \in Q_T$ we have $p(a,t,x) > 0$, then

$$1 \geq \widetilde{h}_N(a,t,x) = \sum_{i=N+1}^{k_N} \lambda_i^N h_i(a,t,x)$$

$$= \sum_{i=N+1}^{k_N} \lambda_i^N \cdot \frac{p_{\varepsilon_i}(a,t,x)}{p_{\varepsilon_i}(a,t,x)+\varepsilon_i} \to 1 = h(a,t,x),$$

as $N \to +\infty$.

If $p(a,t,x) = 0$, then $\widetilde{h}_N(a,t,x) \in [0,1]$, so we conclude that $h(a,t,x) \in [0,1] \subset sgn\ p(a,t,x)$.

It remains to prove that (2.2) has a unique solution.

Let p_1 and p_2 be two solutions to (2.2). It means that there exist $u_1, u_2 \in L^\infty(Q_T)$, $u_i(a,t,x) \in -\rho \cdot sgn\ p_i(a,t,x)$ a.e. $(a,t,x) \in Q_T$, $i \in \{1,2\}$, such that p_i is the solution of (2.2) corresponding to $u := u_i$.

If we denote by $w = p_1 - p_2$, we may conclude that w is the solution of

$$\begin{cases} Dw + \mu w - k\Delta w = u_1 - u_2, & (a,t,x) \in Q_T, \\ \dfrac{\partial w}{\partial \nu}(a,t,x) = 0, & (a,t,x) \in \Sigma_T, \\ w(0,t,x) = \displaystyle\int_0^{a_\dagger} \beta(a) w(a,t,x)da, & (t,x) \in (0,T) \times \Omega, \\ w(a,0,x) = 0, & (a,x) \in (0,a_\dagger) \times \Omega. \end{cases} \tag{2.6}$$

Multiplying the first equation in (2.6) by w and integrating over $(0,a_\dagger) \times (0,t) \times \Omega$, we get that

$$\|w(t)\|^2_{L^2((0,a_\dagger)\times\Omega)} \le \int_0^t \int_\Omega |w(0,s,x)|^2\, dx\, ds$$

(because $(u_1(a,t,x) - u_2(a,t,x))\, w(a,t,x) \le 0$ a.e. $(a,t,x) \in Q_T$)

$$= \int_0^t \int_\Omega \left| \int_0^{a_\dagger} \beta(a) w(a,s,x) da \right|^2 dx\, ds$$

$$\le \|\beta\|^2_{L^2(0,a_\dagger)} \cdot \int_0^t \|w(s)\|^2_{L^2((0,a_\dagger)\times\Omega)}\, ds, \ \forall t \in [0,T].$$

By Bellman's inequality we may infer that

$$w(a,t,x) = 0 \qquad \text{a.e. } (a,t,x) \in Q_T,$$

and we conclude that $p_1(a,t,x) = p_2(a,t,x)$ a.e. $(a,t,x) \in Q_T$ and so the uniqueness holds.

By (2.4) and since $\widetilde{p} \in L^\infty(Q_T)$, we conclude that $p \in L^\infty(Q_T)$.

REMARK 5.2.1. *Using the comparison result for (2.3) and the convergence (2.5), we conclude that*

$$p(a,t,x;p_{01}) \leq p(a,t,x;p_{02}) \qquad a.e.\ (a,t,x) \in Q_T,$$

where p_{01}, p_{02} *satisfy* **(A'5)** *and* $p_{01}(a,x) \leq p_{02}(a,x)$ *a.e.* $(a,x) \in (0,a_\dagger) \times \Omega$. *We have denoted by* $p(\cdot,\cdot,\cdot;p_{0i})$ *the solution of (2.2), corresponding to* $p_0 := p_{0i}$.

We are ready now to formulate the main result of this section:

THEOREM 5.2.2. *If* ρ *is large enough, then the solution* p *to (2.2) satisfies*

$$p(a,T,x) = 0 \qquad \text{a.e. } (a,x) \in (0,a_\dagger) \times \Omega. \tag{2.7}$$

We recall that $p \in C(\overline{S}; L^2(\Omega))$, *for any characteristic line* S, *and so (2.7) is meaningful.*

Proof. Choose

$$M = \|\widetilde{p}\|_{L^\infty((0,a_\dagger)\times\Omega)} \cdot$$

Using Remark 5.2.1 and the comparison result for parabolic equations we get that

$$0 \leq p(a,t,x) \leq \overline{p}(a,t,x) \qquad \text{a.e. in } Q_T, \tag{2.8}$$

where $\overline{p}$ is the solution of

$$\begin{cases} D\overline{p} - k\Delta\overline{p} = -\rho \cdot sgn\ \overline{p}, & (a,t,x) \in Q_T, \\ \dfrac{\partial \overline{p}}{\partial \nu}(a,t,x) = 0, & (a,t,x) \in \Sigma_T, \\ \overline{p}(0,t,x) = M, & (t,x) \in (0,T) \times \Omega, \\ \overline{p}(a,0,x) = M, & (a,x) \in (0,a_\dagger) \times \Omega. \end{cases} \tag{2.9}$$

The solution of (2.9) is

$$\overline{p}(a,t,x) = \begin{cases} M - \rho a, & \text{if } t > a,\ a \leq \dfrac{M}{\rho}, \\ M - \rho t, & \text{if } t < a,\ a \leq \dfrac{M}{\rho}, \\ 0, & \text{otherwise.} \end{cases}$$

So, if $\rho \geq \dfrac{M}{T_0}$ (where $0 < T_0 < a_0, \dfrac{T}{2}$) is arbitrary but fixed, then we conclude that

$$\overline{p}(a,t,x) = 0 \text{ a.e. in } (T_0, a_\dagger) \times (T_0, T) \times \Omega$$

and consequently (by **(A'2)**)

$$\overline{p}(0,t,x) = \int_0^{a_\dagger} \beta(a)\overline{p}(a,t,x)da = 0 \text{ a.e. in } (t,x) \times (T_0,T) \times \Omega.$$

This implies that

$$\overline{p}(a,t,x) = 0 \text{ a.e. in } (a,t,x) \in (0,a_\dagger) \times (T_0,T) \times \Omega$$

and (by (2.8)) consequently

$$p(a,t,x) = 0 \qquad \text{a.e. in } (a,t,x) \in (0,a_\dagger) \times (T_0,T) \times \Omega.$$

We may infer now that

$$p(a,T,x) = 0 \qquad \text{a.e. in } (a,x) \in (0,a_\dagger) \times \Omega.$$

REMARK 5.2.2. *This null controllability result is important mainly because it indicates a feedback control which realizes the null controllability.*

REMARK 5.2.3. *In system (2.1) we can use the feedback control* $u := -\dfrac{1}{T-t}p$, *and the solution* p *corresponding to this control also satisfies*

$$p(a,T,x) = 0, \qquad \text{a.e. } (a,x) \in (0,a_\dagger) \times \Omega,$$

even if the condition **(A'2)** *is not satisfied.*

In fact $\dfrac{1}{T-t}$ *acts as an additional mortality rate. We notice that*

$$\int_0^T \frac{1}{T-t}dt = +\infty,$$

and this condition (which is similar to **(A'4)***) allows us to get the conclusion.*

Actually the first feedback control $(-\rho \cdot \text{sgn } p)$ *is preferable because it belongs to* $L^\infty(Q_T)$.

We leave the details of the proof to the reader.

Consider now the system

$$
\begin{cases}
Dp + \mu(a)p - k\Delta p = f(a,x) + u(a,t,x), & (a,t,x) \in Q_T, \\
\dfrac{\partial p}{\partial \nu}(a,t,x) = 0, & (a,t,x) \in \Sigma_T, \\
p(0,t,x) = \displaystyle\int_0^{a_\dagger} \beta(a)p(a,t,x)da, & (t,x) \in (0,T) \times \Omega, \\
p(a,0,x) = p_0(a,x), & (a,x) \in (0,a_\dagger) \times \Omega,
\end{cases}
\tag{2.10}
$$

where

(A'6) $\quad f \in L^\infty((0,a_\dagger) \times \Omega),\ f(a,x) \geq 0$ a.e. in $(0,a_\dagger) \times \Omega$.

We wish to find a control u such that

$$
p(a,T,x) = p_e(a,x) \qquad \text{a.e. in } (0,a_\dagger) \times \Omega,
$$

where p is the solution to (2.10) and p_e is the solution to

$$
\begin{cases}
p_a + \mu(a)p - k\Delta p = f(a,x), & (a,x) \in (0,a_\dagger) \times \Omega, \\
\dfrac{\partial p}{\partial \nu}(a,t,x) = 0, & (a,x) \in (0,a_\dagger) \times \partial\Omega, \\
p(0,x) = \displaystyle\int_0^{a_\dagger} \beta(a)p(a,x)da, & x \in \Omega.
\end{cases}
$$

Here p_e is a steady state of (2.10), with $u \equiv 0$. This task can be realized by the feedback control

$$
u := -\rho \cdot sgn\ (p - p_e),
$$

with ρ large enough.

We leave the details of the proof to the reader.

REMARK 5.2.4. *The results in this section remain true even in the more general case of vital rates dependent on time and space also. For the sake of simplicity we prefer to treat the case of vital rates depending only on age.*

5.3. A local exact controllability result

This section concerns the local exact controllability of age-dependent population dynamics with diffusion, when the control acts on a subdomain of Ω.

The system we shall deal with is

$$\begin{cases} Dp(a,t,x)+\mu(a)p(a,t,x)-k\Delta p(a,t,x) & = f(a,x) \\ \qquad +m(x)u(a,t,x) & (a,t,x)\in Q_T, \\ \dfrac{\partial p}{\partial \nu}(a,t,x)=0, & (a,t,x)\in \Sigma_T, \\ p(0,t,x)=\displaystyle\int_0^{a_\dagger}\beta(a)p(a,t,x)da, & (t,x)\in(0,T)\times\Omega, \\ p(a,0,x)=p_0(a,x), & (a,x)\in(0,a_\dagger)\times\Omega, \end{cases} \tag{3.1}$$

where u is a control function and m is the characteristic function of ω. Here $\omega\subset\Omega$ is a nonempty open subset.

For the sake of simplicity we assumed that β and μ are time and space independent and the inflow f is time independent.

Actually, we assume that

(A'1) $\beta\in L^\infty(0,a_\dagger)$, $\beta(a)\geq 0$ a.e. $a\in(0,a_\dagger)$;

(A'2) There exists $a_0,a_1\in(0,a_\dagger)$ such that $\beta(a)=0$ a.e. $a\in(0,a_0)\cup(a_1,a_\dagger)$ and $\beta(a)>0$ a.e. in (a_0,a_1);

(A'3) $\mu\in L^1_{loc}([0,a_\dagger))$, $\mu(a)\geq 0$ a.e. $a\in(0,a_\dagger)$;

(A'4) $\displaystyle\int_0^{a_\dagger}\mu(a)da=+\infty$;

(A'5) $p_0\in L^\infty((0,a_\dagger)\times\Omega)$, $p_0(a,t)\geq 0$ a.e. in $(0,a_\dagger)\times\Omega$, $f\in L^\infty((0,a_\dagger)\times\Omega)$, $f(a,x)\geq 0$ a.e. in $(0,a_\dagger)\times\Omega$.

Remark that the assumption **(A'2)** is in accordance with the comments in the first chapter (see also Figure 1).

Let p_e be a steady-state of (3.1), corresponding to $u:=0$ and such that

$$p_e(a,x)\geq\rho_0>0 \qquad \text{a.e. } (a,x)\in(0,a_1)\times\Omega, \tag{3.2}$$

where $\rho_0>0$ is constant.

The main goal of this section is to prove the existence of a control u such that the solution of (3.1) satisfies

$$p(a,T,x)=p_e(a,x) \qquad \text{a.e. } (a,x)\in(0,a_\dagger)\times\Omega$$

and

$$p(a,t,x)\geq 0 \qquad \text{a.e. } (a,t,x)\in Q_T. \tag{3.3}$$

Condition (3.3) is natural because p represents the density of population.

We notice that if p is the solution to (3.1), then $p - p_e$ is the solution to

$$\begin{cases} Dp + \mu(a)p - k\Delta p = m(x)u(a,t,x), & (a,t,x) \in Q_T, \\ \dfrac{\partial p}{\partial \nu}(a,t,x) = 0, & (a,t,x) \in \Sigma_T, \\ p(0,t,x) = \displaystyle\int_0^{a_\dagger} \beta(a)p(a,t,x)da, & (t,x) \in (0,T) \times \Omega, \\ p(a,0,x) = \overline{p}_0(a,x), & (a,x) \in (0,a_\dagger) \times \Omega, \end{cases} \tag{3.4}$$

where $\overline{p}_0 = p_0 - p_e$.

The above formulated problem is equivalent with the exact null controllability problem for (3.4). If we denote now by p the solution to (3.4), then condition (3.3) becomes

$$p(a,t,x) \geq -p_e(a,x) \qquad \text{a.e. } (a,x) \in (0,a_\dagger) \times \Omega,$$

and this shows that the exact null controllability (if this holds) should be local.

The main result of this section amounts to saying that system (3.4) is exactly null controllable for $\overline{p}_0$ in a neighbourhood of 0 (or equivalently for p_0 in a neighbourhood of p_e).

First of all we shall study the existence of p_e, a steady-state of (3.1), corresponding to $u := 0$, which satisfies (3.2). p_e should be a solution to

$$\begin{cases} \dfrac{\partial p_e}{\partial a} + \mu(a)p_e - k\Delta p_e = f(a,x), & (a,x) \in (0,a_\dagger) \times \Omega, \\ \dfrac{\partial p_e}{\partial \nu}(a,x) = 0, & (a,x) \in (0,a_\dagger) \times \partial\Omega, \\ p_e(0,x) = \displaystyle\int_0^{a_\dagger} \beta(a)p_e(a,x)da, & x \in \Omega. \end{cases} \tag{3.5}$$

THEOREM 5.3.1. *If $R < 1$ and $f(a,x) \geq f_0 > 0$ a.e. $(a,x) \in (0,a_\dagger) \times \Omega$, then there exists a unique solution to (3.5), which in addition satisfies (3.2).*

If $R = 1$ and $f \equiv 0$, then there exist infinitely many solutions to (3.5), which satisfy (3.2).

If $R > 1$, then there is no nonnegative solution to (3.5), satisfying (3.2).

Here $R = \displaystyle\int_0^{a_\dagger} \beta(a)e^{-\int_0^a \mu(s)ds}da.$

Proof. If $R < 1$, then there exists a unique (and nonnegative) solution to (3.5) via the Banach fixed point theorem. If in addition $f(a,x) \geq f_0 > 0$ a.e. $(a,x) \in (0,a_\dagger) \times \Omega$, by comparison result in Section 4.1 we get that

$$p_e(a,x) \geq p_i(a,t,x) \qquad \text{a.e. } (a,t,x) \in Q = (0,a_\dagger) \times (0,+\infty) \times \Omega,$$

where p_i is the solution to

$$\begin{cases} Dp_i + \mu p_i - k\Delta p_i = f_0, & (a,t,x) \in Q, \\ \dfrac{\partial p_i}{\partial \nu}(a,t,x) = 0, & (a,t,x) \in \Sigma, \\ p_i(0,t,x) = \displaystyle\int_0^{a_\dagger} \beta(a) p_i(a,t,x)da, & (t,x) \in (0,+\infty) \times \Omega, \\ p_i(a,0,x) = 0, & (a,x) \in (0,a_\dagger) \times \Omega \end{cases}$$

($\Sigma = (0,a_\dagger) \times (0,+\infty) \times \partial\Omega$); p_i does not depend explicitly on x. So, we shall write $p_i(a,t)$ instead of $p_i(a,t,x)$ and

$$p_e(a,x) \geq p_i(a,t) \quad \forall t \in [0,+\infty), \text{ a.e. } (a,x) \in (0,a_\dagger) \times \Omega,$$

where p_i is the solution of

$$\begin{cases} Dp_i + \mu p_i = f_0, & (a,t) \in (0,a_\dagger) \times (0,+\infty), \\ p_i(0,t) = \displaystyle\int_0^{a_\dagger} \beta(a) p_i(a,t)da, & t \in (0,+\infty), \\ p_i(a,0) = 0, & a \in (0,a_\dagger). \end{cases}$$

For $t > a_\dagger$ we have

$$p_i(0,t) > 0 \quad \text{and} \quad t \mapsto p_i(0,t) \text{ is continous}$$

(see Section 2.1). As a consequence we obtain that there exists $\rho_0 > 0$ such that for any t large enough, and for any $a \in (0,a_1)$:

$$p_i(a,t) > \rho_0 \qquad \text{a.e. in } (0,a_1) \times \mathbf{R}^+$$

and in conclusion we get that p_e satisfies (3.2).

If $R = 1$ and $f \equiv 0$, then any function defined by

$$p(a,x) = ce^{-\int_0^a \mu(s)ds}da, \qquad (a,x) \in (0,a_\dagger) \times \Omega$$

is a solution to (3.5) (for any $c \in \mathbf{R}$). In fact these are all the solutions to (3.5) in this case. It is now obvious that there exist infinitely many solutions to (3.5), which satisfy (3.2).

If $R > 1$ and if there exists a nonnegative solution p_e to (3.5) satisfying (3.2), then $p(a,t,x) = p_e(a,x)$, $(a,t,x) \in Q$ is the solution to

$$\begin{cases} Dp + \mu p - k\Delta p = f(a,x), & (a,t,x) \in Q, \\ \dfrac{\partial p}{\partial \nu}(a,t,x) = 0, & (a,t,x) \in \Sigma, \\ p(0,t,x) = \displaystyle\int_0^{a_\dagger} \beta(a)p(a,t,x), & (t,x) \in (0,+\infty) \times \Omega, \\ p(a,0,x) = p_e(a,x), & (a,x) \in (0,a_\dagger) \times \Omega, \end{cases}$$

and for $t \to +\infty$ we have (see Section 4.2)

$$\lim_{t \to +\infty} \|p(t)\|_{L^2((0,a_\dagger)\times\Omega)} = +\infty.$$

On the other hand

$$\|p(t)\|_{L^2((0,a_\dagger)\times\Omega)} = \|p_e\|_{L^2((0,a_\dagger)\times\Omega)},$$

and so $\|p_e\|_{L^2((0,a_\dagger)\times\Omega)} = +\infty$, which is absurd.

LOCAL NULL CONTROLLABILITY FOR (3.4)

THEOREM 5.3.2. *If $\|\overline{p}_0\|_{L^\infty((0,a_\dagger)\times\Omega)}$ is small enough, then there exists $u \in L^2(Q_T)$ such that the solution p of (3.4) satisfies*

$$p(a,T,x) = 0 \quad \text{a.e. } (a,x) \in (0,a_\dagger) \times \Omega$$

and

$$p(a,t,x) \geq -p_e(a,x) \quad \text{a.e. } (a,t,x) \in Q_T.$$

Proof. In what follows we shall use the general Carleman inequality for linear parabolic equations given in [38]. Namely, let $\widetilde{\omega} \subset\subset \omega$ be a nonempty bounded set, $T_0 \in (0,+\infty)$ and $\psi \in C^2(\overline{\Omega})$ be such that

$$\psi(x) > 0, \quad \forall x \in \Omega,$$

$$\psi(x) = 0, \quad \forall x \in \partial\Omega,$$

$$|\nabla\psi(x)| > 0, \quad \forall x \in \overline{\Omega}\backslash\widetilde{\omega}$$

and set

$$\alpha(t,x) = \frac{e^{\lambda\psi(x)} - e^{2\lambda\|\psi\|_{C(\overline{\Omega})}}}{t(T_0 - t)},$$

where λ is an appropriate positive constant.

Define $D_{T_0} = (0, T_0) \times \Omega$.

LEMMA 5.3.3. *There exist positive constants C_1, s_1 such that*

$$\begin{aligned}
&\frac{1}{s}\int_{D_{T_0}} t\,(T_0 - t)\, e^{2s\alpha}\left(|w_t|^2 + |\Delta w|^2\right) dx\ dt + s\int_{D_{T_0}} \frac{e^{2s\alpha}}{t\,(T_0 - t)}\,|\nabla w|^2\, dx\ dt \\
&+s^3\int_{D_{T_0}} \frac{e^{2s\alpha}}{t^3\,(T_0 - t)^3}\,|w|^2\, dx\ dt \qquad (3.6)\\
&\le C_1\left[\int_{D_{T_0}} e^{2s\alpha}\,|w_t + \Delta w|^2\, dx\ dt + s^3\int_{D_{T_0}} \frac{e^{2s\alpha}}{t^3\,(T_0 - t)^3}\,|w|^2\, dx\ dt\right],
\end{aligned}$$

for all $w \in C^2(\overline{D}_{T_0})$, $\frac{\partial w}{\partial\nu}(t,x) = 0$, $\forall (t,x) \in (0,T_0) \times \partial\Omega$ and $s \ge s_1$.

For the proof of this result we refer to [38].

Proof of Theorem 5.3.2 (continued). Let $T_0 \in (0, \min\{a_0, a - a_1\})$. It follows that $T_0 \le T/2$. Define

$$K = L^\infty\left((0, 2T_0) \times \Omega\right).$$

In what follows we shall denote by the same symbol C, several constants independent of $\overline{p}_0$ and all other variables.

For $b \in K$ arbitrary but fixed and for any $\varepsilon > 0$, consider the following optimal control problem $(\mathbf{P}_\varepsilon)$:

$$Minimize\left\{\int_G\int_\Omega \varphi(a,t,x)u^2(a,t,x)dx\ dt\ da + \frac{1}{\varepsilon}\int_{\Gamma_0}\int_\Omega |p_b^u(a,t,x)|^2\, dx\ dl\right\}$$

subject to (3.7) ($u \in L^2(Q_{2T_0})$ and p_b^u is the solution of (3.7) corresponding to u).

$$\begin{cases}
Dp + \mu p - k\Delta p = m(x)\widetilde{m}(a,t)\cdot u(a,t,x), & (a,t,x) \in Q_{2T_0},\\[2mm]
\dfrac{\partial p}{\partial\nu}(a,t,x) = 0, & (a,t,x) \in \Sigma_{2T_0},\\[2mm]
p(0,t,x) = b(t,x), & (t,x) \in (0,2T_0) \times \Omega,\\[2mm]
p(a,0,x) = p_0(a,x), & (a,x) \in (0,a_\dagger) \times \Omega.
\end{cases} \qquad (3.7)$$

Here

$$G = (0, a_\dagger) \times (0, T_0) \cup (0, T_0) \times (T_0, 2T_0),$$

$$\Gamma_0 = \{T_0\} \times (T_0, 2T_0) \cup (T_0, a_\dagger - T_0) \times \{T_0\},$$

$\widetilde{m}$ is the characteristic function of G (see figure 3) and

$$\varphi(a,t,x) = \begin{cases} e^{-2s\alpha(t,x)} t^3 (T_0 - t)^3, & \text{if } t < a, \ (a,t) \in G, \\ e^{-2s\alpha(a,x)} a^3 (T_0 - a)^3, & \text{if } a < t, \ (a,t) \in G. \end{cases}$$

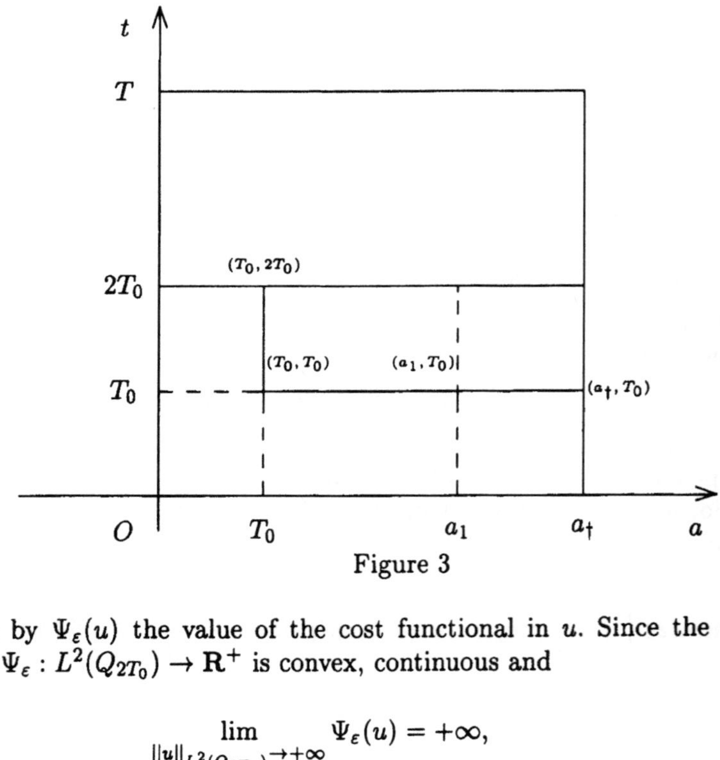

Figure 3

Denote by $\Psi_\varepsilon(u)$ the value of the cost functional in u. Since the cost functional $\Psi_\varepsilon : L^2(Q_{2T_0}) \to \mathbf{R}^+$ is convex, continuous and

$$\lim_{\|u\|_{L^2(Q_{2T_0})} \to +\infty} \Psi_\varepsilon(u) = +\infty,$$

it follows that there exists at least one optimal control u_ε for $(\mathbf{P}_\varepsilon)$ (which is a minimum point for Ψ_ε; see Appendix 1). Let p_ε be the solution of (3.7) corresponding to u_ε. By standard arguments follows

LEMMA 5.3.4 (The maximum principle).

$$u_\varepsilon(a,t,x) = m(x)\widetilde{m}(a,t)q_\varepsilon(a,t,x)\varphi^{-1}(a,t,x) \text{ a.e. } (a,t,x) \in Q_{2T_0}, \tag{3.8}$$

where q_ε *is the solution of*

$$\begin{cases} Dq - \mu q + k\Delta q = 0, & (a,t,x) \in G \times \Omega, \\ \dfrac{\partial q}{\partial \nu}(a,t,x) = 0, & (a,t,x) \in G \times \partial\Omega, \\ q(a,t,x) = 0, & (a,t,x) \in (\Gamma \setminus \Gamma_0) \times \Omega, \\ q(a,t,x) = -\dfrac{1}{\varepsilon} p_\varepsilon(a,t,x), & (a,t,x) \in \Gamma_0 \times \Omega. \end{cases} \tag{3.9}$$

Here $\Gamma = (0,T_0) \times \{2T_0\} \cup \{a_\dagger\} \times (0,T_0) \cup \Gamma_0 \cup (a_\dagger - T_0, a_\dagger) \times \{T_0\}$.

Proof. Since

$$\Psi_\varepsilon(u_\varepsilon) \leq \Psi_\varepsilon(u_\varepsilon + \lambda v),$$

for any $v \in L^2(Q_{2T_0})$ and for any $\lambda \in \mathbf{R}$, we get that

$$\begin{aligned} &\int_G \int_\Omega \varphi(a,t,x) u_\varepsilon^2(a,t,x)dx\ dt\ dt + \frac{1}{\varepsilon}\int_{\Gamma_0}\int_\Omega p_\varepsilon^2(a,t,x)dx\ dl \\ &\leq \int_G \int_\Omega \varphi(a,t,x)|u_\varepsilon(a,t,x) + \lambda v(a,t,x)|^2 dx\ dt\ dt \\ &\quad + \frac{1}{\varepsilon}\int_{\Gamma_0}\int_\Omega |p_\varepsilon(a,t,x) + \lambda z(a,t,x)|^2 dx\ dl. \end{aligned} \tag{3.10}$$

Let z be the solution to

$$\begin{cases} Dz + \mu(a)z - k\Delta z = m(x)\widetilde{m}(a,t)v, & (a,t,x) \in Q_{2T_0}, \\ \dfrac{\partial z}{\partial \nu}(a,t,x) = 0, & (a,t,x) \in \Sigma_{2T_0}, \\ z(0,t,x) = \displaystyle\int_0^{a_\dagger} \beta(a) z(a,t,x)da, & (t,x) \in (0,2T_0) \times \Omega, \\ z(a,0,x) = 0, & (a,x) \in (0,a_\dagger) \times \Omega. \end{cases} \tag{3.11}$$

Relation (3.10) implies that

$$\begin{aligned} &2\int_G \int_\Omega \varphi u_\varepsilon v\ dx\ dt\ da + \lambda \int_G \int_\Omega \varphi v^2 dx\ dt\ da \\ &+ \frac{2}{\varepsilon}\int_{\Gamma_0}\int_\Omega p_\varepsilon z\ dx\ dl + \frac{\lambda}{\varepsilon}\int_{\Gamma_0}\int_\Omega z^2 dx\ dl = 0. \end{aligned}$$

As $\lambda \to 0$ we may conclude that

$$\int_G \int_\Omega \varphi u_\varepsilon v \, dx \, dt \, da + \frac{1}{\varepsilon} \int_{\Gamma_0} \int_\Omega p_\varepsilon z \, dx \, dl = 0, \tag{3.12}$$

for any $v \in L^2(Q_{2T_0})$.

Multiplying the first equation in (3.9) by z and integrating on $G \times \Omega$ we get that

$$\int_G \int_\Omega (Dq_\varepsilon - \mu q_\varepsilon + k\Delta q_\varepsilon) \, z \, dx \, dt \, da = 0.$$

Using now $(3.9)_{2-4}$, (3.7) and (3.11) we obtain after a little calculation that

$$\frac{1}{\varepsilon} \int_{\Gamma_0} \int_\Omega p_\varepsilon z \, dx \, dl + \int_G \int_\Omega q_\varepsilon \, (Dz + \mu z - k\Delta z) \, dx \, dt \, da = 0,$$

and this implies

$$\frac{1}{\varepsilon} \int_{\Gamma_0} \int_\Omega p_\varepsilon z \, dx \, dl + \int_G \int_\Omega m\widetilde{m} q_\varepsilon v \, dx \, dt \, da = 0. \tag{3.13}$$

Relations (3.12) and (3.13) allow us to conclude that

$$\int_{Q_{2T_0}} (\varphi(a,t,x) u_\varepsilon(a,t,x) - m(x)\widetilde{m}(a,t) q_\varepsilon(a,t,x)) \, v(a,t,x) dx \, dt \, da = 0,$$

for any $v \in L^2(Q_{2T_0})$ and consequently (3.8) holds.

Proof of Theorem 5.3.2. (continued). Multiplying the first equation in (3.9) by p_ε and integrating on Q_{2T_0} we obtain after some calculation (and using (3.7) and (3.8)) that

$$\int_G \int_\omega \varphi(a,t,x) |u_\varepsilon(a,t,x)|^2 dx \, da \, dt + \frac{1}{\varepsilon} \int_{\Gamma_0} \int_\Omega |p_\varepsilon(a,t,x)|^2 dx \, dl$$

$$= - \int_0^{T_0} \int_\Omega b(t,x) q_\varepsilon(0,t,x) dx \, dt - \int_0^{a_\dagger - T_0} \int_\Omega \overline{p}_0(a,x) q_\varepsilon(a,0,x) dx \, da.$$

Let S be an arbitrary characteristic line of equation

$$S = \{(\gamma + t, \theta + t); \; t \in (0, T_0)\},$$

$(\gamma, \theta) \in (0, a_\dagger - T_0) \times \{0\} \cup \{0\} \times (0, T_0)$ and define

$$\widetilde{u}_\varepsilon(t,x) = u_\varepsilon(\gamma + t, \theta + t, x),$$

$$\widetilde{p}_\varepsilon(t,x) = p_\varepsilon(\gamma + t, \theta + t, x),$$

$$\widetilde{q}_\varepsilon(t,x) = q_\varepsilon(\gamma+t,\theta+t,x),$$

$(t,x) \in (0,T_0) \times \Omega$,

$$\widetilde{\mu}(t) = \mu(\gamma+t),$$

$t \in (0,T_0)$. $(\widetilde{u}_\varepsilon, \widetilde{p}_\varepsilon, \widetilde{q}_\varepsilon)$ satisfies

$$\begin{cases} (\widetilde{p}_\varepsilon)_t + \widetilde{\mu}\widetilde{p}_\varepsilon - k\Delta\widetilde{p}_\varepsilon = m(x)\widetilde{u}_\varepsilon(t,x), & (t,x) \in (0,T_0)\times\Omega, \\ \dfrac{\partial \widetilde{p}_\varepsilon}{\partial \nu} = 0, & (t,x) \in (0,T_0)\times\partial\Omega, \\ \widetilde{p}_\varepsilon(0,x) = \begin{cases} b(\theta,x), \\ \overline{p}_0(\gamma,x), \end{cases} & \begin{matrix}\gamma=0 \\ \theta = 0\end{matrix} \quad x\in\Omega. \end{cases} \tag{3.14}$$

This yields

$$\widetilde{u}_\varepsilon(t,x) = m(x)\widetilde{q}_\varepsilon(t,x)\cdot e^{2s\alpha(t,x)}t^{-3}(T_0-t)^{-3} \tag{3.15}$$

a.e. $(t,x) \in (0,T_0)\times\Omega$, where

$$\begin{cases} (\widetilde{q}_\varepsilon)_t + k\Delta\widetilde{q}_\varepsilon = \widetilde{\mu}\widetilde{q}_\varepsilon, & (t,x)\in(0,T_0)\times\Omega, \\ \dfrac{\partial\widetilde{q}_\varepsilon}{\partial\nu} = 0, & (t,x)\in(0,T_0)\times\partial\Omega, \\ \widetilde{q}_\varepsilon(T_0,x) = -\dfrac{1}{\varepsilon}\widetilde{p}_\varepsilon(T_0,x) & x\in\Omega. \end{cases} \tag{3.15$'$}$$

Multiplying the first equation by $\widetilde{p}_\varepsilon$ and integrating on D_{T_0} we obtain that

$$\int_0^{T_0}\int_\omega e^{-2s\alpha(t,x)}t^3(T_0-t)^3|\widetilde{u}_\varepsilon(t,x)|^2dx\,dt$$
$$+\frac{1}{\varepsilon}\int_\Omega |\widetilde{p}_\varepsilon(T_0,x)|^2dx = -\int_\Omega \widetilde{p}_\varepsilon(0,x)\widetilde{q}_\varepsilon(0,x)dx. \tag{3.16}$$

By Carleman's inequality (3.6) we infer that

$$\int_0^{T_0}\int_\Omega e^{2s\alpha}[\frac{t(T_0-t)}{s}\left(|(\widetilde{q}_\varepsilon)_t|^2 + |\Delta\widetilde{q}_\varepsilon|^2\right) + \frac{s}{t(T_0-t)}|\nabla\widetilde{q}_\varepsilon|^2$$
$$+\frac{s^3}{t^3(T_0-t)^3}|\widetilde{q}_\varepsilon|^2]dx\,dt$$
$$\leq C_1\int_0^{T_0}\int_\Omega e^{2s\alpha}\,\|\widetilde{\mu}\|^2_{C([0,T_0])}\cdot|\widetilde{q}_\varepsilon|^2\,dx\,dt \tag{3.17}$$
$$+C_1s^3\int_{(0,T_0)\times\omega}\frac{e^{2s\alpha}}{t^3(T_0-t)^3}|\widetilde{q}_\varepsilon|^2\,dx\,dt,$$

and consequently

$$\int_0^{T_0}\int_\Omega e^{2s\alpha}[\frac{t(T_0-t)}{s}\left(|(\widetilde{q}_\varepsilon)_t|^2+|\Delta\widetilde{q}_\varepsilon|^2\right)$$

$$+\frac{s}{t(T_0-t)}|\nabla\widetilde{q}_\varepsilon|^2+\frac{s^3}{t^3(T_0-t)^3}|\widetilde{q}_\varepsilon|^2]dx\,dt$$

$$\leq C\int_0^{T_0}\int_\Omega e^{2s\alpha}\frac{s^3}{t^3(T_0-t)^3}|\widetilde{q}_\varepsilon|^2\,dx\,dt,$$

for $s\geq C\max\{s_1,\|\mu\|^{\frac{2}{3}}_{C([0,a_\dagger-T_0])}\}$.

Multiplying the first equation in (3.15)′ by $\widetilde{q}_\varepsilon$ we obtain that

$$\frac{1}{2}\frac{d}{dt}\int_\Omega|\widetilde{q}_\varepsilon(t,x)|^2\,dx-k\int_\Omega|\nabla_x\widetilde{q}_\varepsilon(t,x)|^2\,dx-\int_\Omega\widetilde{\mu}(t)\,|\widetilde{q}_\varepsilon(t,x)|^2\,dx=0$$

and

$$\frac{d}{dt}\int_\Omega|\widetilde{q}_\varepsilon(t,x)|^2\,dx\geq 0 \text{ a.e. } t\in(0,T_0).$$

Integrating the last inequality we get that

$$\int_\Omega|\widetilde{q}_\varepsilon(0,x)|^2\,dx\leq C\int_{T_0/4}^{3T_0/4}\int_\Omega|\widetilde{q}_\varepsilon(t,x)|^2\,dx,$$

and consequently

$$\int_\Omega|\widetilde{q}_\varepsilon(0,x)|^2\,dx\leq C\int_0^{T_0}\int_\Omega|\widetilde{q}_\varepsilon(t,x)|^2\,\frac{e^{2s\alpha(x,t)}}{t^3\,(T_0-t)^3}dx.$$

By Carleman's inequality we get that

$$\int_\Omega|\widetilde{q}_\varepsilon(0,x)|^2\,dx\leq C\int_0^{T_0}\int_\omega|\widetilde{q}_\varepsilon(t,x)|^2\cdot\frac{e^{2s\alpha(x,t)}}{t^3(T_0-t)^3}dx\,dt.$$

By (3.16) and (3.15) we obtain that

$$\int_{(0,T_0)\times\omega}e^{-2s\alpha}t^3(T_0-t)^3\,|\widetilde{u}_\varepsilon(t,x)|^2\,dx\,dt+\frac{1}{\varepsilon}\int|\widetilde{p}_\varepsilon(T_0,x)|^2\,dx$$

$$\leq C\,\|\widetilde{p}_\varepsilon(0)\|^2_{L^2(\Omega)},$$

for $s\geq C\max\{s_1,\|\mu\|^{\frac{2}{3}}_{C([0,a_\dagger-T_0])}\}$.

Using now (3.17) we conclude that

$$\int_0^{T_0} \int_\Omega \frac{e^{2s\alpha}}{t^3(T_0-t)^3} \left(|\widetilde{q}_\varepsilon(t)|^2 + |\Delta \widetilde{q}_\varepsilon|^2 \right) dx\, dt \le C \|\widetilde{p}_\varepsilon(0)\|^2_{L^2(\Omega)},$$

for any $\varepsilon > 0$ and consequently

$$\|\widetilde{v}_\varepsilon\|^2_{W^2_{1,2}((0,T_0)\times\Omega)} \le C \|\widetilde{p}_\varepsilon(0)\|^2_{L^2(\Omega)},$$

where $\widetilde{v}_\varepsilon(t,x) = e^{2s\alpha(t,x)} t^{-3}(T_0-t)^{-3}\widetilde{q}_\varepsilon(t,x)$, $(t,x) \in (0,T_0)\times\Omega$. As

$$W_2^{1,2}((0,T_0)\times\Omega) \subset L^l((0,T_0)\times\Omega)$$

(where $l = +\infty$ for $n = 1, 2$ and $\frac{1}{l} = \frac{1}{2} - \frac{2}{5}$ for $n = 3$ and consequently $W_2^{1,2}((0,T_0)\times\Omega) \subset L^{10}((0,T_0)\times\Omega)$) we may infer that

$$\|m\widetilde{v}_\varepsilon\|^2_{L^{10}((0,T_0)\times\Omega)} \le C \|\widetilde{p}_\varepsilon(0)\|^2_{L^2(\Omega)}, \tag{3.18}$$

for any $\varepsilon > 0$ and $s \ge C \max\{s_1, \|\mu\|^{\frac{2}{3}}_{C([0,a_\dagger - T_0])}\}$.

The last estimate and the existence theory of parabolic boundary value problems in L^r (see [58], [66]) imply that on a subsequence we have that

$$\widetilde{u}_\varepsilon \to \widetilde{u} \quad \text{weakly in } L^{10}((0,T_0)\times\Omega),$$

$$\widetilde{p}_\varepsilon \to \widetilde{p^{u}} \quad \text{weakly in } W_{10}^{1,2}((0,T_0)\times\Omega),$$

where $\left(\widetilde{u}, \widetilde{p^{u}}\right)$ satisfies (3.14) and

$$\widetilde{p^{u}}(T_0, x) = 0 \quad \text{a.e. } x \in \Omega$$

(with $\tilde{u}_\varepsilon := \tilde{u}$). By (3.14) we get that

$$\left\|\widetilde{p^{u}}\right\|^2_{L^\infty((0,T_0)\times\Omega)} \le C \left(\left\|\widetilde{p^{\tilde{u}}}(0)\right\|^2_{L^\infty(\Omega)} + \|m\widetilde{u}\|^2_{L^3((0,T_0)\times\Omega)} \right)$$

(we recall that $W_3^{1,2}((0,T_0)\times\Omega) \subset L^\infty((0,T_0)\times\Omega)$ for $n \in \{1,2,3\}$; see [1], [66])

$$\le C \left\|\widetilde{p^{\tilde{u}}}(0)\right\|^2_{L^\infty(\Omega)}.$$

For (u, p^u) given by $(\tilde{u}, \tilde{p}^u)$ we have (on each characteristic line) that $u \in L^2(Q_T)$ (u is extended by 0 outside $G \times \Omega$), p^u is the solution of (3.4) and $p(a,t,x) = 0$ a.e. $(a,t,x) \in \Gamma_0 \times \Omega$. Moreover

$$\|p^u\|_{L^\infty(Q_{2T_0})} \le C \left(\|\overline{p}_0\|_{L^\infty((0,a_\dagger)\times\Omega)} + \|b\|_{L^\infty((0,2T_0)\times\Omega)} \right). \tag{3.19}$$

For any $b \in K$, we denote by $\Phi(b) \subset L^\infty((0, 2T_0) \times \Omega)$ the set of all $\int_0^{a_\dagger} \beta(a)p^u(a,t,x)da$, such that $u \in L^2(Q_{2T_0})$, p^u satisfies (3.19) and

$$p^u(a,t,x) = 0, \quad \text{a.e. } (a,t,x) \in \Gamma_0 \times \Omega.$$

There exists an element in $\Phi(b)$ which does not depend on b. Indeed, if $t > T_0$, then

$$\int_0^{a_\dagger} \beta(a)p^u(a,t,x)da = 0$$

because $\beta(a) = 0$ a.e. $a \in (0, T_0)$ and

$$\int_0^{a_\dagger} \beta(a)p^u(a,t,x)da = \int_{T_0}^{a_\dagger - T_0} \beta(a)p^u(a,t,x)da \quad \text{if } t \in (0, T_0)$$

and this depends only on $\overline{p}_0$ and not on b.

In conclusion, $p^u(a, 2T_0, x) = 0$ a.e. $(a,x) \in (0, a_\dagger) \times \Omega$ and

$$\int_{T_0}^{a_\dagger - T_0} \beta(a)p^u(a,t,x)da \le C \|\beta\|_{L^\infty(0,a_\dagger)} \cdot \|\overline{p}_0\|_{L^\infty((0,a_\dagger)\times\Omega)} \cdot \tag{3.20}$$

So, for any (but fixed) u as above we can take

$$b(t,x) = \begin{cases} 0 & \text{a.e. } (t,x) \in (T_0, 2T_0) \times \Omega, \\ \displaystyle\int_0^{a_\dagger} \beta(a)p^u(a,t,x)da & \text{a.e. } (t,x) \in (0, T_0) \times \Omega, \end{cases}$$

which is of course a fixed point of Φ. By (3.19) and (3.20) we get that

$$\|p^u\|_{L^\infty(Q_{2T_0})} \le C_2 \|\overline{p}_0\|_{L^\infty((0,a_\dagger)\times\Omega)}$$

(C_2 is a positive constant).

So, if $\|\overline{p}_0\|_{L^\infty((0,a_\dagger)\times\Omega)}$ is small enough, there exists $u \in L^2(Q_{2T_0})$ and p, the solution of (3.4), satisfies

$$p(a, 2T_0, x) = 0 \quad \text{a.e. } (a,x) \in (0, a_\dagger) \times \Omega$$

and

$$\|p\|_{L^\infty(Q_{2T_0})} \le C_2 \|\overline{p}_0\|_{L^\infty((0,a_\dagger)\times\Omega)} \le \rho_0.$$

In conclusion, $p(a,t,x) \ge -\rho_0$ a.e. $(a,t,x) \in Q_{2T_0}$. This implies that

$$p(a,t,x) \ge -p_e(a,x) \qquad \text{a.e. in } (0, a_\dagger - T_0) \times (0, 2T_0) \times \Omega.$$

On the other hand the control is zero for $(a,t,x) \in (a_\dagger - T_0, a_\dagger) \times (0, 2T_0) \times \Omega$ and so

$$p(a,t,x) \geq -p_e(a,x) \qquad \text{a.e. in } Q_{2T_0}.$$

We conclude the null controllability for (3.4).

It would be important to find a feedback control (as in Section 5.2) which realizes the internal null controllability.

*

* *

The optimal control of age-dependent population dynamics with diffusion is a very important problem and has been investigated by several mathematicians. The optimal harvesting problem for a general nonlinear model has been investigated by B. Ainseba, S. Aniţa and M. Langlais [2]. For related results for some simpler models we refer to Brokate [25], Gurtin and Murphy [46], [47], Murphy and Smith [75]. Some other cost functionals were investigated in [6], [24]. The author discussed the problem with impulsive controllers in [8]; see also [9].

For basic results concerning optimal control problems with distributed parameters we refer to Barbu [16] and Lions [66].

The controllability results for age-dependent population dynamics with diffusion are very recent and were established by B. Ainseba and S. Aniţa [3]. We mention that important results concerning the internal exact controllability problem for the linear heat equation were published in 1995 by G. Lebeau and L. Robbiano [62].

The approach in Section 5.1 is inspired by the paper of B. Ainseba, S. Aniţa and M. Langlais [2].

The results in Section 5.2 were obtained by the author.

Section 5.3 is devoted to the internal exact controllability and is based on the paper of B. Ainseba and S. Aniţa [3].

5.4. Exercises

Consider the optimal control problem

$$\textbf{(E1)} \qquad \textit{Minimize} \int_0^1 \int_0^1 \int_0^1 |p^u(a,t,x) - 1|^2 dx\, da\, dt,$$

subject to $u \in L^\infty((0,1)\times(0,1)\times(0,1))$, $0 \le u(a,t,x) \le L$ $(L>0)$, where p^u is the solution to

$$
\begin{aligned}
& Dp + p - p_{xx} = u(a,x,t), && (a,t,x) \in (0,1)\times(0,1)\times(0,1),\\
& p_x(a,t,0) = p_x(a,t,1) = 0, && (a,t) \in (0,1)\times(0,1),\\
& p(0,t,x) = \int_0^1 p(a,t,x)da, && (t,x) \in (0,1)\times(0,1),\\
& p(a,0,x) = \cos \pi x, && (a,x) \in (0,1)\times(0,1).
\end{aligned}
$$

1. Prove that problem **(E1)** admits at least one optimal pair and find the optimality conditions.

   Hint. Since the cost function I_1 is a lower semicontinuous convex and proper function, and since in addition

$$\lim_{\|u\|_{L^2((0,1)\times(0,1)\times(0,1))}\to+\infty} I_1(u) = +\infty,$$

   we may conclude via Theorem A.1.2 that I_1 admits at least one minimum point. The deduction of the optimality conditions follows in the same manner as in Section 5.1.

Consider the optimal harvesting problem

$$\textbf{(E2)} \qquad \textit{Maximize} \int_0^{2\pi}\int_0^1\int_0^1 u(a,t,x)p^u(a,t,x)dx\, da\, dt,$$

subject to $u \in L^\infty((0,1)\times(0,2\pi)\times(0,1))$, $0 \le u(a,t,x) \le L$ $(L>0)$, where p^u is the solution to

$$
\begin{aligned}
& Dp + \frac{1}{1-a}p - p_{xx} = 1 - \cos t \\
& \quad -u(a,x,t)p(a,t,x), && (a,t,x) \in (0,1)\times\mathbf{R}\times(0,1),\\
& p_x(a,t,0) = p_x(a,t,1) = 0, && (a,t) \in (0,1)\times\mathbf{R},\\
& p(0,t,x) = \int_0^1 p(a,t,x)da, && (t,x) \in \mathbf{R}\times(0,1),\\
& p(a,t,x) = p(a,t+2\pi,x), && (a,t,x) \in (0,1)\times\mathbf{R}\times(0,1).
\end{aligned}
$$

2. Prove that there exists an optimal control u^* for problem **(E2)**.

   Hint. The proof follows the same steps as the proof of Theorem 5.1.2.

3. Show that u^* (the optimal control for **(E2)**) satisfies

$$u^*(a,t,x) = \begin{cases} 0, & \text{if } 1+q(a,t,x)<0, \\ L, & \text{if } 1+q(a,t,x)>0, \end{cases} \tag{4.1}$$

   where q is the solution to

$$\begin{aligned}
&Dq - \frac{1}{1-a}q + q_{xx} = -q(0,t,x) && +L(1+q(a,t,x))^+, \\
& && (a,t,x) \in (0,1)\times \mathbf{R} \times (0,1), \\
&q_x(a,t,0) = q_x(a,t,1) = 0, && (a,t) \in (0,1) \times \mathbf{R}, \\
&q(1,t,x) = 0, && (t,x) \in \mathbf{R} \times (0,1), \\
&q(a,t,x) = q(a,t+2\pi,x), && (a,t,x) \in (0,1)\times \mathbf{R} \times (0,1).
\end{aligned} \tag{4.2}$$

   Hint. The existence of the solution to (4.2) follows by a similar argument to that in Section 2.4. Condition (4.1) follows in the same manner as the main result in Section 5.1.

4. Show that u^* is a bang-bang control.

5. Assume that the hypotheses in Section 5.3 hold and that $\omega \neq \Omega$. Prove that there is no control $u \in L^2((0,a_\dagger)\times(0,T)\times\omega)$ such that

$$p(a,t,x) \geq 0 \quad \text{a.e. in } Q_T,$$

   and

$$p(a,T,x) = 0 \quad \text{a.e. in } (0,a_\dagger)\times\Omega,$$

   where p is the solution to (3.1) corresponding to $p_0 \equiv 1$.

   Hint. If such a control would exist, then the solution p of (3.1) satisfies

$$p(a,t,x) \geq h(a,t,x) \quad \text{a.e. in } (0,a_\dagger)\times(0,T)\times(\Omega\setminus\omega),$$

where h satisfies

$$\begin{cases} Dh + \mu(a)h - k\Delta h = 0 & (a,t,x) \in (0,a_\dagger) \times (0,T) \times (\Omega \setminus \omega), \\ \frac{\partial h}{\partial \nu}(a,t,x) = 0, & (a,t,x) \in \Sigma_T, \\ h(a,t,x) = 0, & (a,t,x) \in (0,a_\dagger) \times (0,T) \times \partial\omega, \\ h(0,t,x) = \int_0^{a_\dagger} \beta(a)h\, da, & (t,x) \in (0,T) \times (\Omega \setminus \omega), \\ h(a,0,x) = 1, & (a,x) \in (0,a_\dagger) \times (\Omega \setminus \omega). \end{cases}$$

Since $h(a,T,x) = 0$ a.e. in $(0,a_\dagger) \times (\Omega \setminus \omega)$, it follows that $h(a,0,x) = 0$ a.e in $(0,a_\dagger) \times (\Omega \setminus \omega)$; which is false.

Assume that the hypotheses in Section 5.3 hold and that in addition

$$R = \int_0^{a_\dagger} \beta(a) e^{-\int_0^a \mu(s)ds} da = 1,$$

$$p_0 \in L^\infty((0,a_\dagger) \times \Omega).$$

Consider the problem

$$\begin{cases} Dp + \mu(a)p - k\Delta p \in -m(x) \cdot sgn\, p(a,t,x), & \\ & (a,t,x) \in (0,a_\dagger) \times (0,+\infty) \times \Omega, \\ \frac{\partial p}{\partial \nu}(a,t,x) = 0, & (a,t,x) \in (0,a_\dagger) \times (0,+\infty) \times \partial\Omega, \\ p(0,t,x) = \int_0^{a_\dagger} \beta(a)p\, da, & (t,x) \in (0,+\infty) \times \Omega, \\ p(a,0,x) = p_0(a,x), & (a,x) \in (0,a_\dagger) \times \Omega. \end{cases}$$

6. Prove that

$$\lim_{t \to +\infty} \|p(t)\|_{L^\infty((0,a_\dagger)\times\Omega)} = 0.$$

APPENDIX 1: ELEMENTS OF NONLINEAR ANALYSIS

A1.1. Convex functions and subdifferentials

Throughout this section X will be a real Banach space with the norm $\|\cdot\|$ and dual X^*. We shall denote by $(\cdot,\cdot)$ the pairing between X and X^*. If X is a Hilbert space identified with its dual, then $(\cdot,\cdot)$ is the scalar product of X.

The function $\varphi : X \to (-\infty, +\infty]$ is called *convex* if

$$\varphi(\lambda x + (1-\lambda)y) \leq \lambda\varphi(x) + (1-\lambda)\varphi(y), \tag{1.1}$$

for all $x, y \in X$ and $\lambda \in [0,1]$. If the inequality (1.1) is strict for $x \neq y$ and $\lambda \in (0,1)$, then the function φ is called *strictly convex.*

The set

$$D(\varphi) = \{x \in X;\ \varphi(x) < +\infty\}$$

is called *the effective domain of* φ and

$$epi\ (\varphi) = \{(x,\lambda) \in X \times \mathbf{R};\ \varphi(x) \leq \lambda\}$$

is the *epigraph of* φ.

The function $\varphi : X \to \overline{\mathbf{R}}$ is called *lower semicontinuous at* $x_0 \in X$ if

$$\liminf_{x \to x_0} \varphi(x) \geq \varphi(x_0).$$

A function $\varphi : X \to \overline{\mathbf{R}}$ is said to be *lower semicontinuous on* X (or simply *lower semicontinuous*) if it is lower semicontinuous at every point $x_0 \in X$.

It is readily seen that the function φ is convex if and only if *epi* (φ) is a convex subset of $X \times \mathbf{R}$.

On the other hand φ is lower semicontinuous on X if and only if *epi* (φ) is closed in $X \times \mathbf{R}$.

Note also that φ is lower semicontinuous on X if and only if every level set $\{x \in X;\ \varphi(x) \leq \lambda\}$ $(\lambda \in \mathbf{R})$ is closed in X.

The function $\varphi : X \to (-\infty, +\infty]$ is said to be *proper* if $\varphi \not\equiv +\infty$.

As a consequence of the Hahn-Banach separation theorem we obtain that a convex set of X is closed if and only if it is weakly closed (see Brézis [23]).

THEOREM A1.1.1. *The convex function* $\varphi : X \to (-\infty, +\infty]$ *is lower semicontinuous on* X *if and only if it is weakly lower semicontinuous on* X, *i.e.,*

$$\liminf_{x \rightharpoonup x_0} \varphi(x) \geq \varphi(x_0), \quad \forall x_0 \in X.$$

Proof. Since the level set $\{x \in X;\ \varphi(x) \leq \lambda\}$ (which is convex) is closed if and only if it is weakly closed, we get the conclusion of the theorem.

THEOREM A1.1.2. *Let* X *be a reflexive Banach space and let* $\varphi : X \to (-\infty, +\infty]$ *be a lower semicontinuous convex function. If* X_0 *is a closed, convex and bounded subset of* X, *then* φ *attains its infimum on* X_0. *In other words, there exists* $x_0 \in X_0$ *such that*

$$\varphi(x_0) = \inf\{\varphi(x);\ x \in X_0\}.$$

Proof. Let $\{x_n\}_{n \in \mathbf{N}^*} \subset X_0$ be such that

$$\varphi(x_n) \to \inf\{\varphi(x);\ x \in X_0\}$$

as $n \to +\infty$. Since X_0 is a closed, convex and bounded subset of the reflexive Banach space X, it is compact in the weak topology of X. In conclusion there exists a subsequence, also denoted by $\{x_n\}$ such that $x_n \to x_0$ weakly in X. By Theorem A1.1.1. we infer that φ is weakly lower semicontinuous on X and in conclusion we get that

$$\liminf_{n \to +\infty} \varphi(x_n) \geq \varphi(x_0)$$

and consequently

$$\varphi(x_0) = \inf\{\varphi(x);\ x \in X_0\}.$$

As a consequence we get

THEOREM A1.1.3. *Let* X *be a reflexive Banach space and let* $\varphi : X \to (-\infty, +\infty]$ *be a lower semicontinuous convex function such that*

$$\lim_{\|x\| \to +\infty} \varphi(x) = +\infty.$$

Then φ *attains its infimum on* X.

THEOREM A1.1.4. *Let $\varphi : X \to (-\infty, +\infty]$ be a lower semicontinuous and convex function. Then φ is continuous on the interior of $D(\varphi)$.*

Proof. Let $x_0 \in int\ D(\varphi)$. Without loss of generality we may assume that $x_0 = 0$, $\varphi(0) = 0$. To prove that φ is continuous in 0 it suffices to show that for every $\lambda > 0$ the level set $X_\lambda = \{x \in X;\ \varphi(x) \leq \lambda\}$ is a neighbourhood of the origin. The set

$$E = X_\lambda \cap \{x \in X;\ \varphi(-x) \leq \lambda\}$$

is closed, convex, symmetric and absorbs every point of X. By Baire's category theorem, E is a neighbourhood of the origin and consequently X_λ is a neighbourhood of the origin.

THEOREM A1.1.5. *Let $\varphi : X \to (-\infty, +\infty]$ be a convex, lower semicontinuous and proper function. Then φ is bounded from below by an affine function on X.*

Proof. Let $x_0 \in D(\varphi)$. For any $\varepsilon > 0$ it follows that $(x_0, \varphi(x_0) - \varepsilon) \notin$ *epi* (φ). By the Hahn-Banach separation theorem (see Brézis [23]) there exists a linear continuous functional on $X \times \mathbf{R}$ ($(x^*, \lambda) \in (X^* \times \mathbf{R})$) such that

$$(x, x^*) + \lambda t \leq \gamma < (x_0, x^*) + \lambda(\varphi(x_0) - \varepsilon),$$

for any $(x, t) \in$ *epi* (φ) (here γ is a real constant). For $x \in D(\varphi)$ and $t = \varphi(x)$ we get $\lambda < 0$.

This yields

$$\varphi(x) \geq \varphi(x_0) - \varepsilon + \frac{1}{\lambda}(x_0 - x, x^*),$$

for any $x \in X$ and we get the conclusion of the theorem.

Let K be a closed convex subset of X. The function $I_K : X \to (-\infty, +\infty]$,

$$I_K(x) = \begin{cases} 0, & \text{if } x \in K, \\ +\infty, & \text{if } x \in X \setminus K \end{cases}$$

is called *the indicator function of* K. This function is convex and lower semicontinuous on X.

Let $\varphi : X \to (-\infty, +\infty]$ be a lower semicontinuous convex function. The *subdifferential of* φ is the multivalued function $\partial\varphi : X \to X^*$ defined by

$$\partial\varphi(x) = \{x^* \in X^*;\ (y - x, x^*) \leq \varphi(y) - \varphi(x) \text{ for all } y \in X\}.$$

The elements of $\partial\varphi(x)$ are called *subgradients* of φ at x.

It is obvious that $\partial\varphi(x)$ is closed and convex for any $x \in X$.

The set $D(\partial\varphi) = \{x \in X;\ \partial\varphi(x) \neq \emptyset\}$ is called the *domain of* $\partial\varphi$.

The definition of $\partial\varphi$ implies that x is a minimum point of φ if and only if

$$0 \in \partial\varphi(x).$$

Let K be a closed convex subset of X. The set

$$N_K(x) = \partial I_K(x)$$

is called *the normal cone to K at x*.

The function $f : X \to \mathbf{R}$ is *Gâteaux differentiable at* $x_0 \in X$ if

$$\lim_{\varepsilon \to 0+} \varepsilon^{-1}\left(f(x_0 + \varepsilon x) - f(x_0)\right) = h(x)$$

exists for any $x \in X$ and h is a linear continuous functional on X. Denote by $\nabla f(x_0) \in X^*$ (the gradient of f at x_0) the element defined by

$$h(x) = (x, \nabla f(x_0)) \qquad \text{for all } x \in X.$$

THEOREM A1.1.6. *Let $\varphi : X \to (-\infty, +\infty]$ be a lower semicontinuous convex function. Then $int\ D(\varphi) \subset D(\partial\varphi)$.*

Proof. Let $x_0 \in int\ D(\varphi)$. By Theorem A1.1.4, φ is continuous at x_0 and we conclude that $(x_0, \varphi(x_0) + \varepsilon) \in int\ epi\ (\varphi)$ for any $\varepsilon > 0$. The sets $\{(x_0, \varphi(x_0))\}$ and $epi\ (\varphi)$ can be separated by a closed hyperplane. It is obvious that the hyperplane passes through $(x_0, \varphi(x_0))$. It follows that there exists $x^* \in X^*$ and $\lambda \geq 0$ such that

$$(x - x_0, x^*) \leq \lambda(\varphi(x) - \varphi(x_0)),$$

for any $x \in D(\varphi)$.

Since $(x^*, \lambda) \neq (0, 0)$, we may conclude that λ should be strictly positive. So, we may infer that $\lambda^{-1}x^* \in \partial\varphi(x_0)$.

APPROXIMATION OF CONVEX FUNCTIONS

Assume that X is a Hilbert space and let $\varphi : X \to (-\infty, +\infty]$ be a convex, lower semicontinuous and proper function. For any $\varepsilon > 0$ we define the function $\varphi_\varepsilon \to \mathbf{R}$,

$$\varphi_\varepsilon(x) = \inf\left\{\frac{\|x - y\|^2}{2\varepsilon} + \varphi(y);\ y \in X\right\}. \tag{1.2}$$

By Theorem A1.1.3 we conclude that for every $x \in X$ and $\varepsilon > 0$, the infimum in (1.2) is attained. As the infimum of a family of lower semicontinuous convex functions is a lower semicontinuous and convex function, we may conclude that φ_ε is a lower semicontinuous and convex function on X.

THEOREM A1.1.7 (Moreau - Brézis). *The function φ_ε is Fréchet differentiable on X. Moreover, we have*

$$\varphi_\varepsilon(x) = \varepsilon\|\nabla\varphi_\varepsilon\|^2/2 + \varphi(J_\varepsilon x),$$

for all $x \in X$;

$$\lim_{\varepsilon\to 0+} \varphi_\varepsilon(x) = \varphi(x),$$

for all $x \in X$;

$$\varphi(J_\varepsilon x) \le \varphi_\varepsilon(x) \le \varphi(x),$$

for all $x \in X$.

Here we have denoted by

$$J_\varepsilon = (I + \varepsilon\partial\varphi)^{-1}.$$

For the proof we refer to [21].

EXAMPLE A1.1.1

Let Ω be a measurable bounded subset of $\mathbf{R}^n$ ($n \in \mathbf{N}^*$) and $f : \Omega \times \mathbf{R} \to \overline{\mathbf{R}}$ satisfying the assumptions:

(1) f is measurable on $\Omega \times \mathbf{R}$, i.e., it is measurable with respect to the σ-algebra of subsets of $\Omega \times \mathbf{R}$ generated by products of Lebesgue measurable subsets of Ω and Borelian subsets of $\mathbf{R}$;

(2) $f(x,\cdot) : \mathbf{R} \to \overline{\mathbf{R}}$ is a convex, lower semicontinuous and proper function for almost any $x \in \Omega$;

(3) $f(x,t) \ge a_1(x)t + a_2(x)$, for almost any $(x,t) \in \Omega \times \mathbf{R}$, where $a_1 \in L^2(\Omega)$ and $a_2 \in L^1(\Omega)$;

(4) there exists $y_0 \in L^2(\Omega)$ such that $f(\cdot, y_0) \in L^1(\Omega)$.

A function f which satisfies (1) and (2) is called a *normal convex integrand on $\Omega \times \mathbf{R}$*. If in addition f does not depend on the first variable, then (3) and (4) are satisfied.

Define the function $I_f : L^2(\Omega) \to (-\infty, +\infty]$,

$$I_f(y) = \begin{cases} \displaystyle\int_\Omega f(x, y(x))dx & \text{if } f(\cdot, y(\cdot)) \in L^1(\Omega), \\ +\infty & \text{otherwise.} \end{cases}$$

By assumptions (1) and (2) we deduce that the function $x \mapsto f(x, y(x))$ is Lebesgue measurable on Ω for any measurable function (on Ω) y.

THEOREM A1.1.8. *Under the hypotheses* (1) – (4), *the function* I_f *is convex, lower semicontinuous and proper.*

Moreover, for every $y \in X$ *we have*

$$\partial I_f(y) = \{v \in L^2(\Omega);\ v(x) \in \partial_y f(x, y(x)) \text{ a.e. } x \in \Omega\},$$

where $\partial_y f$ *is the subdifferential as a function of* y.

For the proof we refer to [21].

EXAMPLE A1.1.2

Let Ω be a measurable bounded subset of $\mathbf{R}^n$ ($n \in \mathbf{N}^*$) and $\zeta_1, \zeta_2 \in L^2(\Omega)$, with $\zeta_1(x) \le \zeta_2(x)$ a.e. $x \in \Omega$.

Define

$$K = \{y \in L^2(\Omega);\ \zeta_1(x) \le y(x) \le \zeta_2(x) \text{ a.e. } x \in \Omega\}.$$

This is a closed convex subset of $L^2(\Omega)$.

We shall prove that

$$N_K(y) = \{z \in L^2(\Omega);\ z(x) \le 0 \text{ for } y(x) = \zeta_1(x) < \zeta_2(x),$$
$$\text{and } z(x) \ge 0 \text{ for } y(x) = \zeta_2(x) > \zeta_1(x)\},$$

for any $y \in K$.

Indeed, since $N_K(y) = \partial I_K(y)$, for any $y \in K$, it will be our goal to characterize the elements of $\partial I_K(y)$. An element $z \in L^2(\Omega)$ belongs to $\partial I_K(y)$ if and only if

$$\int_\Omega z(x)\,(v(x) - y(x))\,dx \le 0,$$

for any $v \in K$.

For any $z \in \partial I_K(y)$ we denote by $\Omega^1 = \{x \in \Omega;\ y(x) = \zeta_1(x) < \zeta_2(x)\}$ and we take $v \in K$ such that $v(x) = y(x)$ for almost any $x \in \Omega \setminus \Omega^1$. We conclude that

$$\int_{\Omega^1} z(x)\,(v(x) - \zeta_1(x))\,dx \le 0,$$

for any $v \in L^2(\Omega^1)$, $\zeta_1(x) \leq v(x) \leq \zeta_2(x)$ a.e. $x \in \Omega^1$. In conclusion $z(x) \leq 0$ a.e. in Ω^1. In the same manner it follows that $z(x) \geq 0$ for almost any $x \in \Omega$ such that $y(x) = \zeta_2(x) > \zeta_1(x)$.

Conversely, it is obvious that any $z \in L^2(\Omega)$ such that $z(x) \leq 0$ for $y(x) = \zeta_1(x)$ and $z(x) \geq 0$ for $y(x) = \zeta_2(x)$, belongs to $\partial I_K(y)$.

REMARK. For a complete treatment of the subject in this section we refer to Rockafellar [77].

A1.2. Generalized gradients of locally Lipschitz functions

We shall present the Clarke generalized gradient and some of its properties. This section is based on Clarke's papers [31], [32]. Let X be a real Banach space with the norm $\|\cdot\|$ and dual X^*. We shall denote by $(\cdot,\cdot)$ the pairing between X and X^*.

Let $f : X \to \mathbf{R}$ be a *locally Lipschitz function*, i.e., for every $r > 0$, there exists $L(r) > 0$ such that

$$\|f(x) - f(y)\| \leq L(r)\|x - y\|,$$

for any $x, y \in B(0; r) = \{z \in X;\ \|z\| \leq r\}$.

We define the *generalized directional derivative of f at $x \in X$ in the direction $h \in X$* to be

$$f^0(x,h) = \limsup_{y \to x, \varepsilon \to 0+} \varepsilon^{-1} \left(f(y + \varepsilon h) - f(y)\right). \tag{2.1}$$

It is obvious that $f^0(x,h) \in \mathbf{R}$ for any $(x,h) \in X \times X$ and that the function $h \mapsto f^0(x,h)$ is subadditive, positively homogeneous. In addition, for any $x \in X$, there exists $C > 0$ such that

$$f^0(x,h) \leq C\|h\|,$$

for any $h \in X$.

By the Hahn-Banach theorem we may infer that there exists at least one $\xi \in X^*$ such that

$$(h,\xi) \leq f^0(x,h), \tag{2.2}$$

for all $h \in X$.

The set of all $\xi \in X^*$ satisfying (2.2) is called *the generalized gradient (Clarke's gradient) of f at x*. It is denoted by $\partial f(x)$.

For any locally Lipschitz function $f : X \to \mathbf{R}$, the function $f^0 : X \times X \to \mathbf{R}$ defined by (2.1) is upper semicontinuous. Indeed, let $\{x_n\}_{n \in \mathbf{N}^*}$,

$\{h_n\}_{n\in\mathbf{N}^*} \subset X$ be strongly convergent to x and h, respectively. For every $n \in \mathbf{N}^*$ there exists $y_n \in X$ and $\varepsilon_n \in (0,1)$ such that $\|y_n\| + \varepsilon_n \leq \dfrac{1}{n}$ and

$$f^0(x_n, h_n) \leq \varepsilon_n^{-1} \left(f(x_n + y_n + \varepsilon_n h_n) - f(x_n + y_n)\right) + \frac{1}{n}.$$

Letting $n \to +\infty$ we get that

$$\limsup_{n\to+\infty} f^0(x_n, h_n) \leq f^0(x, h)$$

and we get the conclusion.

THEOREM A1.2.1. *Let $f : X \to \mathbf{R}$ be a locally Lipschitz function. Then for every $x \in X$, $\partial f(x)$ is a convex and weak star compact subset of X^*.*

In addition ∂f is weak star upper semicontinuous, i.e.,

"If $\{x_n\}_{n\in\mathbf{N}^*} \subset X$, $x_n \to x$ strongly in X and $\{x_n^*\}_{n\in\mathbf{N}^*} \subset X^*$, $x_n^* \in \partial f(x_n)$ $(n \in \mathbf{N}^*)$, $x_n^* \to x^*$ weak star in X^*, then $x^* \in \partial f(x)$."

Proof. By (2.1) and (2.2) it follows that for any $x \in X$, $\partial f(x)$ is convex, closed and bounded. As a consequence $\partial f(x)$ is weak star compact.

Since f^0 is upper semicontinuous we may conclude that ∂f is weak star semicontinuous.

The following fundamental results hold for $f : X \to \mathbf{R}$ a locally Lipschitz function:

1. If f is convex, then $\partial f(x)$ coincides with the subdifferential of f at x;
2. If f attains a local extremum at x, then $0 \in \partial f(x)$;
3. If f admits a continuous Gâteaux derivative ∇f, then $\partial f(x) = \nabla f(x)$.

For the proofs of these results we refer to [32].

AN APPROXIMATING RESULT

Assume that X is a real separable Hilbert space and let $\{e_n\}_{n\in\mathbf{N}^*}$ be an orthonormal basis on X. Consider X_n the finite dimensional subspace generated by $\{e_1, e_2, \ldots, e_n\}$ $(n \in \mathbf{N}^*)$.

Denote by $P_n : X \to X_n$ the projection of X on X_n. This yields

$$P_n x = \sum_{i=1}^{n} x_i e_i,$$

where $x = \sum_{i=1}^{\infty} x_i e_i$.

Consider the operator $\Lambda_n : \mathbf{R}^n \to X_n$ defined by

$$\Lambda_n \theta = \sum_{i=1}^{n} \theta_i e_i,$$

where $\theta = (\theta_1, \theta_2, ..., \theta_n)$.

For the locally Lipschitz function $f : X \to \mathbf{R}$ and for any $\varepsilon > 0$ we define $f_\varepsilon : X \to \mathbf{R}$,

$$f_\varepsilon(x) = \int_{\mathbf{R}^n} f(P_n x - \varepsilon \Lambda_n \theta) \rho_n(\theta) d\theta, \quad x \in X.$$

Here $n = \left[\frac{1}{\varepsilon}\right]$ and $\rho_n \in C_0^\infty(\mathbf{R}^n)$ is a C^∞-mollifier on $\mathbf{R}^n$, i.e.,

$$\rho_n(\theta) = 0, \quad \text{for } \|\theta\|_{\mathbf{R}^n} > 1, \quad \rho_n(\theta) \geq 0,$$

for any $\theta \in \mathbf{R}^n$ and

$$\int_{\mathbf{R}^n} \rho_n(\theta) d\theta = 1, \qquad \rho_n(\theta) = \rho_n(-\theta),$$

for all $\theta \in \mathbf{R}^n$.

THEOREM A1.2.2. *The function f_ε is continuously Fréchet differentiable. In addition*

$$\lim_{\varepsilon \to 0^+} f_\varepsilon(x) = f(x), \tag{2.3}$$

for all $x \in X$.

If $x_\varepsilon \to x$ in X (for $\varepsilon \to 0^+$) and if $\nabla f_\varepsilon(x_\varepsilon) \to \eta$ weakly in X, then

$$\eta \in \partial f(x).$$

Proof. Using the substitution $\theta = \varepsilon^{-1}(\Lambda_n^{-1} P_n x - \tau)$ (in the definition formula for f_ε) we immediately obtain that

$$f_\varepsilon(x) = \frac{1}{\varepsilon^n} \int_{\mathbf{R}^n} f(\Lambda_n \tau) \rho_n(\varepsilon^{-1}(\Lambda^{-1} P_n x - \tau)) d\tau,$$

for all $x \in X$.

Since $\rho_n \in C_0^\infty(\mathbf{R}^n)$ we may conclude that f_ε is Fréchet differentiable and that ∇f_ε is continuous on X.

Using again the definition of f_ε we may conclude that

$$f_\varepsilon(x) - f(x) = \int_{\mathbf{R}^n} \left(f(P_n x - \varepsilon \Lambda_n \theta) - f(x)\right) \rho_n(\theta) d\theta$$

(because $\int_{\mathbf{R}^n} \rho_n(\tau)d\tau = 1$) and that for every $x \in X$, there exists $C > 0$ such that

$$\|f_\varepsilon(x) - f(x)\| \leq C\left(\|x - P_n x\| + \varepsilon\right).$$

The last inequality implies that (2.3) holds.

If we assume now that $x_\varepsilon \to x$ in X (for $\varepsilon \to 0^+$) and $\nabla f_\varepsilon(x_\varepsilon) \to \eta$ weakly in X, then we may apply the mean value theorem and we get that

$$\lambda^{-1}\left(f_\varepsilon(x+\lambda y) - f_\varepsilon(x_\varepsilon)\right)$$

$$= \lambda^{-1} \sum_{i=0}^{n} \gamma_{i,n,\lambda}(f(P_n(x+\lambda y) - \varepsilon\Lambda_n \xi_{i,n,\lambda}) - f(P_n x_\varepsilon - \varepsilon \Lambda_n \xi_{i,n,\lambda}). \tag{2.4}$$

Here $\gamma_{i,n,\lambda} \in [0,1]$, $\sum_{i=0}^{n} \gamma_{i,n,\lambda} = 1$, $\|\xi_{i,n,\lambda}\|_{\mathbf{R}^n} \leq 1$. We may extract two subsequences (also denoted by $\{\gamma_{i,n,\lambda}\}$ and $\{\xi_{i,n,\lambda}\}$) such that

$$\gamma_{i,n,\lambda} \to \gamma_{i,n}, \qquad \xi_{i,n,\lambda} \to \xi_{i,n},$$

for $\lambda \to 0$. By (2.4) we obtain that

$$(\nabla f_\varepsilon(x_\varepsilon), P_n y) \leq \sum_{i=0}^{n} \gamma_{i,n} f^0(P_n x - \varepsilon \Lambda_n \xi_{i,n}, P_n y),$$

for any $y \in X$ and since f^0 is upper semicontinuous we conclude that

$$\eta \in \partial f(x).$$

A1.3. The Ekeland variational principle

This section concerns a very important result in nonconvex optimization. It is known in the mathematical literature as the Ekeland variational principle (see Ekeland [35], [36]). The result is very useful if we wish to prove the existence of a minimum point in the absence of standard compactness conditions.

Let (X, d) be a complete metric space and $f : X \to (-\infty, +\infty]$ be a lower semicontinuous function, bounded from below and nonidentically $+\infty$.

THEOREM A1.3.1. *Let $\varepsilon > 0$ and let $u \in X$ be such that*

$$f(u) \leq \inf\{f(x);\ x \in X\} + \varepsilon.$$

Then, for any $\lambda > 0$ there exists $u_\varepsilon \in X$ such that

$$f(u_\varepsilon) \leq f(u), \tag{3.1}$$

$$d(u_\varepsilon, u) \le \lambda, \tag{3.2}$$

$$f(x) > f(u_\varepsilon) - \varepsilon\lambda^{-1}d(u_\varepsilon, x), \quad \text{for any } x \in X \setminus \{u_\varepsilon\}. \tag{3.3}$$

Proof. We shall first prove the case $\lambda = 1$. We define inductively the sequence $\{u_n\}_{n\in\mathbf{N}}$ in the following manner: $u_0 = u$ and if u_{n-1} is known, then we define
$u_n = u_{n-1}$ if

$$f(x) > f(u_{n-1}) - \varepsilon d(u_{n-1}, x) \quad \text{for any } x \in X \setminus \{u_{n-1}\}.$$

Otherwise we consider the set $A_n = \{x \in X;\ f(x) \le f(u_{n-1}) - \varepsilon d(u_{n-1}, x)\}$ and choose $u_n \in A_n$ such that

$$f(u_n) - \inf\{f(x);\ x \in A_n\} \le \frac{1}{2}\left(f(u_{n-1}) - \inf\{f(x);\ x \in A_n\}\right). \tag{3.4}$$

We shall prove that $\{u_n\}_{n\in\mathbf{N}}$ is convergent.

The definition of the sequence implies that

$$\varepsilon d(u_k, u_n) \le f(u_k) - f(u_n), \tag{3.5}$$

for any $k,\ n \in \mathbf{N}$, $k \le n$. Since $\{f(u_n)\}_{n\in\mathbf{N}}$ is a decreasing sequence and bounded from below, we may infer that it is a convergent sequence and consequently it is a Cauchy sequence. By (3.5) we deduce that $\{u_n\}_{n\in\mathbf{N}}$ is a Cauchy sequence and since (X, d) is a complete metric space we conclude that it is convergent. Denote by u_ε the limit of $\{u_n\}_{n\in\mathbf{N}}$.

Since $f(u) \ge f(u_n)$, for any $n \in \mathbf{N}$ and by using the convergence of $\{u_n\}_{n\in\mathbf{N}}$ and the lower semicontinuouity of f we get that (3.1) holds.

By (3.5) we get that

$$\varepsilon d(u, u_n) \le f(u) - f(u_n) \le f(u) - \inf\{f(x);\ x \in X\} \le \varepsilon,$$

for any $n \in \mathbf{N}$ and in conclusion we obtain (3.2).

If there exists $x \in X \setminus \{u_\varepsilon\}$ such that

$$f(x) \le f(u_\varepsilon) - \varepsilon d(u_\varepsilon, x),$$

then we shall get a contradiction. Indeed, since $f(u_\varepsilon) \le f(u_n)$, we infer that

$$f(x) \le f(u_n) - \varepsilon d(u_n, x)$$

and consequently $x \in A_{n+1}$ for any $n \in \mathbf{N}$. By (3.4) it follows that

$$2f(u_{n+1}) - f(u_n) \le f(x)$$

and $f(x) \geq f(u_\varepsilon)$, which is absurd. This implies that (3.3) holds.

In the case of a general $\lambda > 0$, we define the metric $d_1 = \lambda^{-1}d$. If we apply the already proved result (for $\lambda = 1$) for the complete metric space (X, d_1) and the function f, we get the conclusion of the theorem.

If in addition X is a Banach space endowed with the norm $\|\cdot\|$ and f is in addition Gâteaux differentiable, then the following result holds:

THEOREM A1.3.2. *For any $\varepsilon > 0$ there exists $u_\varepsilon \in X$ such that*

$$f(u_\varepsilon) \leq \inf \{f(x);\ x \in X\} + \varepsilon,$$

$$\|\nabla f(u_\varepsilon)\| \leq \varepsilon^{\frac{1}{2}}.$$

The proof follows immediately by Theorem A1.3.1.

APPENDIX 2: THE LINEAR HEAT EQUATION

We shall recall some results in the theory of linear parabolic equations. For more details we refer to Barbu [17].

In order to discuss the concept of weak solution for a linear parabolic equation we need to present some notions and results related to infinite dimensional vectorial functions on real intervals. Let X be a Banach space with the norm $\|\cdot\|$.

Let $a,\ b \in \mathbf{R}$, $a < b$. The function $f : [a,b] \to X$ is said to be *continuous at* $t_0 \in [a,b]$ if the real function $g : [a,b] \to \mathbf{R}$,

$$g(t) = \|f(t) - f(t_0)\|, \qquad t \in [a,b]$$

is continuous at t_0. If f is continuous at each $t \in [a,b]$, we say that f is *continuous on* $[a,b]$.

Denote by $C([a,b];X)$ the space of all continuous functions $f : [a,b] \to X$. This is a Banach space with the norm

$$\|f\|_{C([a,b];X)} = \max\{\|f(t)\|;\ t \in [a,b]\}.$$

The function $f : [a,b] \to X$ is said to be *absolutely continuous* on $[a,b]$ if for each $\varepsilon > 0$ there exists $\delta(\varepsilon) > 0$ such that

$$\sum_{k=1}^{N} \|f(t_k) - f(s_k)\| < \varepsilon,$$

whenever $\sum_{k=1}^{N} |t_k - s_k| < \delta(\varepsilon)$ and $(t_k, s_k) \cap (t_j, s_j) = \emptyset$, for $j \neq k$ ($t_k,\ s_k \in [a,b]$). Denote by $AC([a,b];X)$ the space of all absolutely continuous functions $f : [a,b] \to X$.

The function $f : [a,b] \to X$ is said to be *differentiable at* $t_0 \in [a,b]$ if there exists $f'(t_0) \in X$ (the derivative of f at t_0) such that

$$\lim_{t \to t_0} \left\| \frac{1}{t - t_0}(f(t) - f(t_0)) - f'(t_0) \right\| = 0.$$

The derivative f' is also denoted by $\dfrac{df}{dt}$. We say that f is *differentiable on* $[a,b]$ if f is differentiable at each $t \in [a,b]$.

By $C^1([a,b];X)$ we shall denote the space of all differentiable functions $f : [a,b] \to X$ with continuous derivative f' on $[a,b]$.

THEOREM A2.1. *Let X be a reflexive Banach space. If $f \in AC([a,b];X)$, then f is almost everywhere differentiable on $[a,b]$ and*

$$f(t) = f(a) + \int_a^t \frac{df}{dt}(s)ds, \quad \forall t \in [a,b].$$

For the proof we refer to [15].

A function $f : [a,b] \to X$ is said to be *finitely valued* if there exist a finite number of disjoint measurable subsets $A_k \subset [a,b]$, $k \in \{1,2,...,N\}$, such that $[a,b] = \cup_{k=1}^N A_k$ and f is constant on each A_k. The function $f : (a,b) \to X$ is *measurable* if there exists a sequence f_n of finite valued functions such that

$$f_n(t) \to f(t)$$

almost everywhere on (a,b).

Denote by $L^p(a,b;X)$ ($p \in [1,+\infty]$) the space of all measurable functions $f : (a,b) \to X$ such that the function

$$t \mapsto \|f(t)\| = g(t)$$

belongs to $L^p(a,b)$. The usual norm on $L^p(a,b;X)$ is

$$\|f\|_{L^p(a,b;X)} = \|g\|_{L^p(a,b)}.$$

We can now complete Theorem A2.1 by

THEOREM A2.2. *If $f \in C([a,b];X)$ and $\frac{df}{dt} \in L^1(a,b;X)$, then*

$$f \in AC([a,b];X).$$

For the proof we refer to [17].

Let X be a Hilbert space with the scalar product denoted $(\cdot,\cdot)$ and the norm $\|\cdot\|$. The system $\{\varphi_j\}_{j=1}^\infty \subset X$ is called *orthonormal* if

$$(\varphi_j, \varphi_k) = \delta_{jk}, \ \forall j,k \in \mathbf{N}^*,$$

where δ_{jk} is Kronecker's symbol.

The orthonormal system is said to be *complete* (orthonormal basis) if it is maximal in the class of all orthonormal systems, i.e., if $\varphi \in X$ satisfies $(\varphi, \varphi_j) = 0$ for all j, this implies that $\varphi = 0$.

LEMMA A2.3. *Let $\{\varphi_j\}_{j=1}^{\infty}$ be an orthonormal system. Then the following conditions are equivalent:*

(i) $\displaystyle\sum_{j=1}^{\infty} |(\varphi, \varphi_j)|^2 = \|\varphi\|^2, \ \forall \varphi \in X$ *(the Parseval formula).*

(ii) $\displaystyle\varphi = \sum_{j=1}^{\infty} (\varphi, \varphi_j)\varphi_j, \ \forall \varphi \in X.$

(iii) $\{\varphi_j\}_{j=1}^{\infty}$ *is complete.*

Proof. Let $\varphi \in X$ be arbitrary but fixed. Since the system $\{\varphi_j\}_{j=1}^{\infty}$ is orthonormal we conclude that

$$\|\varphi - \sum_{j=1}^{n} (\varphi, \varphi_j)\varphi_j\|^2 = \|\varphi\|^2 - \sum_{j=1}^{n} |(\varphi, \varphi_j)|^2, \ \forall n \in \mathbf{N}^*.$$

This equality implies that conditions (i) and (ii) are equivalent.

If (ii) holds and if $\varphi \in X$ satisfies

$$(\varphi, \varphi_j) = 0 \qquad \forall j \in \mathbf{N}^*,$$

then we have that $\varphi = 0$.

Conversely, if we assume that the system $\{\varphi_j\}$ is complete, then the Cauchy criterion allows us to conclude that the series $\displaystyle\sum_{j=1}^{\infty} |(\varphi, \varphi_j)|^2$ and $\displaystyle\sum_{j=1}^{\infty} (\varphi, \varphi_j)\varphi_j$ are convergent. If we now take

$$\tilde{\varphi} = \varphi - \sum_{j=1}^{\infty} (\varphi, \varphi_j)\varphi_j,$$

then we get that $(\tilde{\varphi}, \varphi_j) = 0$, $\forall j \in \mathbf{N}^*$ and this implies that (ii) holds.

In what follows we shall deal with applications to the case $X = L^2(\Omega)$ or $X = H^k(\Omega)$.

The series $\sum_{j=1}^{\infty} c_j(t)$ where $c_j \in C([a,b]; X)$ is said to be *uniformly convergent* to the function $c : [a,b] \to X$ if

$$\lim_{N \to +\infty} \|\sum_{j=1}^{N} c_j - c\|_{C([a,b];X)} = 0.$$

The following Cauchy criterion holds:

LEMMA A2.4. *The series* $\sum_{j=1}^{\infty} c_j(t)$ *is uniformly convergent if and only if* $\forall \varepsilon > 0$, $\exists N(\varepsilon) \in \mathbf{N}^*$, *such that* $\|\sum_{j=k}^{p} c_j(t)\| < \varepsilon$ *for all* $p > k \geq N(\varepsilon)$, $t \in [a, b]$.

If $\{\varphi_j\}_{j=1}^{\infty}$ is an orthonormal basis in X, then the series $\sum\limits_{j=1}^{\infty} c^j(t)\varphi_j$ ($c^j \in C([a, b])$) is called a *Fourier series*. By Parseval's formula it follows that

LEMMA A2.5. *The Fourier series* $\sum_{j=1}^{\infty} c^j(t)\varphi_j$ *is uniformly convergent on* $[a, b]$ *if and only if the series* $\sum_{j=1}^{\infty}(c^j)^2(t)$ *is uniformly convergent.*

If $\sum_{j=1}^{\infty} \|c^j\|^2_{L^2(a,b)}$ *is convergent, then the series* $\sum_{j=1}^{\infty} c^j\varphi$ *is convergent in* $L^2(a, b; X)$.

We also have

LEMMA A2.6. *If* $c^j \in W^{1,2}(a, b)$, $\sum\limits_{j=1}^{\infty} c^j(t)\varphi_j = c(t)$ *uniformly on* $[a, b]$ ($c \in C([a, b]; X)$) *and* $\sum_{j=1}^{\infty} \|(c^j)'\|^2_{L^2(a,b)}$ *is convergent, then* $c \in AC([a, b]; X)$.

Proof. Define

$$h = \sum_{j=1}^{\infty} (c^j)'\varphi_j$$

in $L^2(a, b; X)$. This yields

$$\int_a^t h(s)ds = \int_a^t \sum_{j=1}^{\infty} (c^j)'(s)\varphi_j ds$$

$$= \sum_{j=1}^{\infty} \left(\int_a^t (c^j)'(s)ds \right) \varphi_j = \sum_{j=1}^{\infty} \left(c^j(t) - c^j(a)\right) \varphi_j = c(t) - c(a),$$

$\forall t \in [a, b]$. This implies that there exists $\frac{dc}{dt}(t) = h(t)$ a.e. $t \in [a, b]$ and that $c \in AC([a, b]; X)$.

Let $\Omega \subset \mathbf{R}^n$ ($n \in \mathbf{N}^*$) be a nonempty open and bounded set with a smooth enough boundary $\partial\Omega$.

Consider the linear parabolic equation

$$\begin{cases} \dfrac{\partial y}{\partial t}(t,x) - k\Delta y(t,x) = g(t,x), & (t,x) \in (0,T_0) \times \Omega, \\ \\ \dfrac{\partial y}{\partial \nu}(t,x) = 0, & (t,x) \in (0,T_0) \times \partial\Omega, \\ \\ y(0,x) = y_0(x), & x \in \Omega, \end{cases} \tag{1}$$

where $T_0, k \in (0, +\infty)$.

The following result holds (see [17]):

LEMMA A2.7. *If $y_0 \in L^2(\Omega)$ and $g \in L^2((0,T_0) \times \Omega)$, then there exists a unique $y \in C([0,T_0]; L^2(\Omega)) \cap AC((0,T_0); L^2(\Omega)) \cap L^2(0,T_0; H^1(\Omega))$ $\cap L^2_{loc}((0,T_0); H^2(\Omega))$, such that*

$$\begin{cases} \dfrac{\partial y}{\partial t}(t,x) - k\Delta y(t,x) = g(t,x), & \text{a.e. in } (0,T_0) \times \Omega, \\ \\ \dfrac{\partial y}{\partial \nu}(t,x) = 0, & \text{a.e. in } (0,T_0) \times \partial\Omega, \\ \\ y(0,x) = y_0(x), & \text{a.e. in } \Omega. \end{cases} \tag{1$'$}$$

If $y_0(x) \geq 0$ a.e. in Ω and $g(t,x) \geq 0$ a.e. in $(0,T_0) \times \Omega$, then $y(t,x) \geq 0$ a.e. $(t,x) \in (0,T_0) \times \Omega$.

If $y_0(x) \geq 0$ a.e. in Ω and $g(t,x) > 0$ a.e. in $(0,T_0) \times \Omega$, then $y(t,x) > 0$ a.e. $(t,x) \in (0,T_0) \times \Omega$.

Moreover, if $\{y_{0N}\}_{N \in \mathbb{N}} \subset L^2(\Omega)$, $y_{0N} \to y_0$ in $L^2(\Omega)$, as $N \to +\infty$, then

$$y_N^0 \to y \quad \text{in } C([0,T_0]; L^2(\Omega)),$$

where y_N^0 is the solution of (1) corresponding to $y_0 := y_{0N}$.

Here $\dfrac{\partial y}{\partial \nu}(t,x) = \nabla_x y(t,x) \cdot \nu_x$, where $\nabla_x y(\cdot,x)$, $x \in \partial\Omega$ is the "trace" of $\nabla_x y(\cdot,x)$ on $\partial\Omega$ and ν_x is the unit outward normal to $\partial\Omega$ at x.

A function $y \in C([0,T_0]; L^2(\Omega)) \cap AC((0,T_0); L^2(\Omega)) \cap L^2(0,T_0; H^1(\Omega))$ $\cap L^2_{loc}((0,T_0); H^2(\Omega))$, which satisfies (1)$'$ is called a *solution* to (1). Here

$$AC((0,T_0); L^2(\Omega)) = \{h : (0,T_0) \to L^2(\Omega);\ h \text{ is absolutely continuous on any compact subinterval}\}.$$

The complete proof of this result can be found in [17]. We are particularly interested in the constructive part of this proof. We will show how the solution of (1) can be constructed as a Fourier series and that this solution belongs to

$$C([0,T_0];L^2(\Omega)) \cap AC((0,T_0);L^2(\Omega)) \cap L^2(0,T_0;H^1(\Omega)).$$

By Theorem 3.6.1 in [17] we know that there is an orthonormal complete system (orthonormal basis) $\{\varphi_j\}_{j\in\mathbf{N}} \subset L^2(\Omega)$ and $\{\lambda_j\} \subset \mathbf{R}^+$, $\lambda_0 = 0$, $\lambda_j \nearrow +\infty$ as $j \to \infty$ such that

$$\begin{cases} -\Delta\varphi_j(x) = \lambda_j\varphi_j(x) & \text{in } \Omega, \\ \dfrac{\partial\varphi_j}{\partial\nu}(x) = 0 & \text{on } \partial\Omega. \end{cases} \tag{2}$$

Recall that $\varphi_j \in H^1(\Omega)$ is a solution to (2) if

$$\int_\Omega \nabla\varphi_j(x)\cdot\nabla\varphi(x)\,dx = \lambda_j\int_\Omega \varphi_j(x)\varphi(x)\,dx, \quad \forall\varphi\in H^1(\Omega).$$

This yields

$$\varphi_0(x) = \frac{1}{(meas(\Omega))^{\frac{1}{2}}}.$$

We look for a solution y of the form

$$y(t,x) = \sum_{j=0}^{\infty} y^j(t)\varphi_j(x), \quad (t,x)\in(0,T_0)\times\Omega, \tag{3}$$

where y^j are unknown functions. If one formally substitutes (3) into (1) we obtain for y^j the differential system

$$\begin{cases} (y^j)'(t) + \lambda_j k y^j(t) = g_j(t), & t\in(0,T_0), \\ y^j(0) = y_0^j, & j\in\mathbf{N}, \end{cases} \tag{4}$$

where

$$\begin{aligned} y_0^j &= (y_0,\varphi_j) = \int_\Omega y_0(x)\varphi_j(x)dx, \\ g_j(t) &= (g(t),\varphi_j) = \int_\Omega g(t,x)\varphi_j(x)dx. \end{aligned} \tag{5}$$

Here $(\cdot,\cdot)$ denotes the usual scalar product in $L^2(\Omega)$. Solving the Cauchy problem (4), we obtain

$$y^j(t) = e^{-\lambda_j kt}y_0^j + \int_0^t e^{-\lambda_j k(t-s)}g_j(s)ds, \quad t\in[0,T_0] \tag{6}$$

and by (3) we obtain that

$$y(t,x) = \sum_{j=0}^{\infty} \left(e^{-\lambda_j kt} y_0^j + \int_0^t e^{-\lambda_j k(t-s)} g_j(s) ds \right) \varphi_j(x). \tag{7}$$

Let us first prove that $y \in C([0,T_0];L^2(\Omega))$. It suffices to show that the series

$$\sum_{j=0}^{\infty} \left(e^{-\lambda_j kt} y_0^j + \int_0^t e^{-\lambda_j k(t-s)} g_j(s) ds \right)^2 \tag{8}$$

is uniformly convergent on $[0,T_0]$ (see Lemma A2.5). Indeed, we have

$$(y^j)^2(t) = \left(e^{-\lambda_j kt} y_0^j + \int_0^t e^{-\lambda_j k(t-s)} g_j(s) ds \right)^2 \leq 2e^{-2\lambda_j kt}(y_0^j)^2$$

$$+2\left(\int_0^t e^{-\lambda_j k(t-s)} g_j(s) ds \right)^2 \leq 2(y_0^j)^2 + 2\int_0^{T_0} |g_j(s)|^2 ds, \ \forall j \in \mathbf{N}, \ t \in [0,T_0].$$

Using now (5), (6) and Parseval's formula we get that

$$\sum_{j=0}^{\infty} (y_0^j)^2 = \|y_0\|^2_{L^2(\Omega)}, \quad \sum_{j=0}^{\infty} |g_j(t)|^2 = \|g(t)\|^2_{L^2(\Omega)}, \ \forall t \in [0,T_0],$$

while by the Beppo-Levi theorem, we have

$$\sum_{j=0}^{\infty} \int_0^{T_0} |g_j(s)|^2 ds = \int_0^{T_0} \left(\sum_{j=0}^{\infty} |g_j(s)|^2 \right) ds = \int_{(0,T_0)\times\Omega} |g(t,x)|^2 dx\, dt.$$

This implies that the series (8) is uniformly convergent on $[0,T_0]$. We shall prove now that $y \in L^2(0,T_0;H^1(\Omega))$. By (7) we may rewrite y as

$$y(t,x) = \sum_{j=0}^{\infty} \sqrt{1+\lambda_j}\, y^j(t) \psi_j(x), \quad (t,x) \in (0,T_0) \times \Omega,$$

where $\psi_j = (\sqrt{1+\lambda_j})^{-1}\varphi_j$. Since $\{\psi_j\}$ is an orthonormal complete system in $H^1(\Omega)$ (this can be proved by direct verification and using (2)), it suffices to show that the series

$$\sum_{j=0}^{\infty} (1+\lambda_j) \|y^j\|^2_{L^2(0,T_0)} \tag{9}$$

is convergent, or equivalently that the series

$$\sum_{j=0}^{\infty} \lambda_j \|y^j\|^2_{L^2(0,T_0)}$$

is convergent (because $\sum_{j=0}^{\infty} \|y^j\|^2_{L^2(0,T_0)} = \|y\|^2_{L^2(0,T_0;L^2(\Omega))}$). This yields (by (8))

$$\lambda_j (y^j)^2(t) \le 2\lambda_j e^{-2\lambda_j kt}(y_0^j)^2 + \frac{1}{k}\int_0^{T_0} |g_j(s)|^2 ds, \quad \forall j \in \mathbf{N}$$

and (by Parseval's formula) consequently

$$\sum_{j=0}^{\infty} \lambda_j \|y^j\|^2_{L^2(0,T_0)} \le \frac{1}{k}\left(\|y_0\|^2_{L^2(\Omega)} + \|g\|^2_{L^2((0,T_0)\times\Omega)}\right).$$

It follows that (9) holds and so $y \in L^2(0,T_0;H^1(\Omega))$.

In order to prove that $y_t \in L^2(\delta, T_0-\delta; L^2(\Omega))$ (for any $\delta > 0$ small enough) it suffices to show that the series $\sum_{j=0}^{\infty} \|y_j'\|^2_{L^2(\delta,T_0-\delta)}$ is convergent for any $\delta > 0$ small enough. Since

$$(y_j)'(t) = -\lambda_j k y^j(t) + g_j(t),$$

and since

$$\sum_{j=0}^{\infty} \|g_j\|^2_{L^2(0,T_0)} = \|g\|^2_{L^2((0,T_0)\times\Omega)}$$

(by Parseval's formula), it suffices to prove the convergence of the series

$$\sum_{j=0}^{\infty} \lambda_j^2 k^2 \|y^j\|^2_{L^2(\delta,T_0-\delta)},$$

for any $\delta > 0$ small enough. By (6) we obtain

$$\lambda_j^2 k^2 (y^j)^2(t) \le 2\lambda_j^2 k^2 e^{-2\lambda_j kt}(y_0^j)^2$$
$$+2\lambda_j^2 k^2 e^{-2\lambda_j kt}\left(\int_0^t e^{\lambda_j ks} ds\right)\cdot\left(\int_0^t e^{\lambda_j ks}|g_j(s)|^2 ds\right)$$
$$\le \frac{2}{t^2}(y_0^j)^2 + 2\lambda_j k e^{-\lambda_j kt}\int_0^t e^{\lambda_j ks}|g_j(s)|^2 ds.$$

Since the series

$$\sum_{j=0}^{\infty}\int_{\delta}^{T_0-\delta} \frac{2}{t^2}(y_0^j)^2 dt$$

is convergent and

$$\sum_{j=0}^{\infty}\int_{\delta}^{T_0-\delta}\lambda_j k e^{-\lambda_j k t}\int_0^t e^{\lambda_j k s}|g_j(s)|^2 ds\, dt$$

$$\leq 2\int_0^{T_0} e^{\lambda_j k s}|g_j(s)|^2\int_s^{T_0}(-e^{-\lambda_j k t})' dt\, ds \leq 2\|g\|^2_{L^2((0,T_0)\times\Omega)},$$

we may conclude that

$$\sum_{j=0}^{\infty}\|(y^j)'\|^2_{L^2(\delta,T-0-\delta)}$$

is convergent. It follows that $y_t \in L^2(\delta, T_0-\delta; L^2(\Omega))$ and that

$$y \in AC((0,T_0); L^2(\Omega)).$$

For the complete proof of Lemma A2.7 we refer to [17].

REFERENCES

[1] Adams, R.A., *Sobolev Spaces*, Academic Press, New York (1975).

[2] Ainseba, B., Aniţa, S., and Langlais, M., "On the optimal harvesting of a nonlinear age-structured population dynamics", *Comm. Appl. Anal.*, to appear.

[3] Ainseba, B., and Aniţa, S., "Local exact controllability of the age dependent population dynamics with diffusion", *Diff. Integral Eqs.*, to appear.

[4] Anderson, R., and May, R., "Vaccination against rubella and measles: quantitative investigations of different policies", *J. Hyg. Camb.* **90** (1983), 259–325.

[5] Aniţa, S., "Approximating linear optimal control problems via Trotter formula", *Nonlin. Anal.; T.M.A.* **12** (1988), 375–388.

[6] Aniţa, S., "Optimal control of a nonlinear population dynamics with diffusion", *J. Math. Anal. Appl.* **152** (1990), 176–208.

[7] Aniţa, S., "Existence and uniqueness for the population dynamics with nonlinear diffusion", *Diff. Integral Eqs.* **4** (1991), 835–850.

[8] Aniţa, S., "Optimal impulse-control of population dynamics with diffusion", in *Differential Equations and Control Theory*, V. Barbu (Ed.), Pitman Research Notes in Mathematics Series, Vol. 250, Longman Scientific and Technical (1992), 1–6.

[9] Aniţa, S., "Optimal control of parameter distributed systems with impulses", *Appl. Math. Optim.* **29** (1994), 93–107.

[10] Aniţa, S., "Optimal harvesting for a nonlinear age-dependent population dynamics", *J. Math. Anal. Appl.* **226** (1998), 6–22.

[11] Aniţa, S., Iannelli, M., Kim, M.-Y., and Park, E.-J., "Optimal harvesting for periodic age-dependent population dynamics", *SIAM J. Appl. Math.* **58** (1998), 1648–1666.

[12] Arnăutu, V., "Numerical results for a product formula approximation of Hamilton-Jacobi equations", *International J. Comp. Math.* **57** (1995), 75–82.

[13] Aronson, D.G., "The role of diffusion in mathematical population biology: Skellam revisited", in *Proc. Internat. Conf. Bari-Italy-1983*, Lecture Notes in Biomathematics (1985), 2–6.

[14] Barbu, V., "A product formula approach to nonlinear optimal control problems", *SIAM J. Control Optim.* **26** (1988), 497–520.

[15] Barbu, V., *Analysis and Control of Infinite Dimensional Systems*, Aca-

demic Press, New-York (1993).

[16] Barbu, V., *Mathematical Methods in Optimization of Differential Systems*, Kluwer Academic Publishers, Dordrecht (1994).

[17] Barbu, V., *Partial Differential Equations and Boundary Value Problems*, Kluwer Academic Publishers, Dordrecht (1998).

[18] Barbu, V., and Iannelli, M., "Approximating some non-linear equations by a fractional step scheme", *Diff. Integral Eqs.* **6** (1993), 15–26.

[19] Barbu, V., and Iannelli, M., "Optimal control of population dynamics", *JOTA* **102** (1999), 1–14.

[20] Barbu, V., and Iannelli, M., "The semigroup approach to non-linear age-structured equations", *Rend. Ist. Mat. Univ. Trieste* **28** (1996), 59–71.

[21] Barbu, V., and Precupanu, T., *Convexity and Optimization in Banach Spaces*, D. Reidel Publishing Company, Dordrecht-Boston (1986).

[22] Bardos, C. and Tartar, L., "Sur l'unicité rétrograde des équations d'évolution", *C.R. Acad. Sci. Paris, Ser. A* **273** (1971), 1239–1241.

[23] Brézis, H., *Analyse Fonctionelle. Théorie et Applications*, Dunod, Paris (1983).

[24] Brokate, M., "Pontryagin's principle for control problems in age dependent population dynamics", *J. Math. Biol.* **23** (1985), 75–101.

[25] Brokate, M., "On a certain optimal harvesting problem with continuous age structure", in *Optimal Control of Partial Differential Equations II*, Birkhäuser (1987), 29–42.

[26] Busenberg, S., and Iannelli, M., "A class of nonlinear diffusion problems in age-dependent population dynamics", *J. Nonlin. Anal.; T.M.A.* **7** (1983), 501–529.

[27] Busenberg, S., and Iannelli, M., "Separable models in age-dependent population dynamics", *J. Math. Biol.* **22** (1985), 145–173.

[28] Chan, W.L., and Guo, B.Z., "Optimal birth control of population dynamics", *J. Math. Anal Appl.* **144** (1989), 532–552.

[29] Chan, W.L., and Guo, B.Z., "Optimal birth control of population dynamics. II. Problems with free final time, phase constraints, and minimax costs", *J. Math. Anal Appl.* **146** (1990), 523–539.

[30] Charlesworth, B., *Evolution in Age-Structured Populations*, Cambridge University Press, Cambridge (1980).

[31] Clarke, F.H., "Generalized gradients and applications", *Trans. Amer. Math. Soc.* **205** (1975), 247–262.

[32] Clarke, F.H., "Generalized gradients and Lipschitz functionals", *Adv. in Math.* **40** (1981), 52–67.

[33] Dietz, K., and Schenzle, D., "Proportionate mixing models for age-dependent infection transmission", *J. Math. Biol.* **22** (1985), 117–120.

[34] Doetsch, G., *Introduction to the Theory and Applications of the Laplace Transformation*, Springer-Verlag, Berlin (1974).

[35] Ekeland, I., "On the variational principle", *J. Math. Anal. Appl.* **47**

(1974), 324–353.

[36] Ekeland, I., "Nonconvex minimization problems", *Bull. Amer. Math. Soc.* **1** (1979), 443–474.

[37] Feller, W., "On the integral equation of renewal theory", *Ann. Math. Stat.* **12** (1941), 243–267.

[38] Fursikov, A.V, and Imanuvilov, O.Yu., *Controllability of Evolution Equations*, Lecture Notes Series 34, RIM Seoul National University, Korea (1996).

[39] Garroni, M.G., and Langlais, M., "Age-dependent population diffusion with external constraint, *J. Math. Biol.* **14** (1982), 77–94.

[40] Gurtin, M.E., "A system of equations for age-dependent population diffusion", *J. Theor. Biol.* **40** (1973), 389–392.

[41] Gurtin, M.E., and MacCamy, R.C., "Nonlinear age-dependent population dynamics", *Arch. Rat. Mech. Anal.* **54** (1974), 281–300.

[42] Gurtin, M.E., and MacCamy, R.C., "On the diffusion of biological population", *Math. Biosci.* **38** (1977), 35–49.

[43] Gurtin, M.E., and MacCamy, R.C., "Some simple models for nonlinear age-dependent population dynamics", *Math. Biosci.* **43** (1979), 199–211.

[44] Gurtin, M.E., and MacCamy, R.C., "Diffusion models for age structured populations", *Math. Biosci.* **54** (1981), 49–59.

[45] Gurtin, M.E., and MacCamy, R.C., "Product solutions and asymptotic behaviour in age-dependent population diffusion", *Math. Biosci.* **62** (1982), 157–167.

[46] Gurtin, M.E., and Murphy, L.F., "On the optimal harvesting of age structured populations: some simple models", *Math. Biosci.* **55** (1981), 115–136.

[47] Gurtin, M.E., and Murphy, L.F., "On the optimal harvesting of persistent age-structured populations", *J. Math. Biol.* **13** (1981), 131–148.

[48] Hadeler, K.P., and Mueller, J., "Vaccination in age-structured populations I: The reproduction number", in *Models for Infectious Human Deseases: Their Structure and Relation to Data*, V. Isham et al (Eds.), Cambridge University Press (1996), 90–101.

[49] Hadeler, K.P., and Mueller, J., "Vaccination in age-structured populations II: Optimal vaccination strategies", in *Models for Infectious Human Deseases: Their Structure and Relation to Data*, V. Isham et al (Eds.), Cambridge University Press (1996), 102–106.

[50] Hoppensteadt, F., *Mathematical Theories of Populations: Demographics, Genetics and Epidemics*, CBMS Lectures 20, SIAM Publ., Philadelphia (1975).

[51] Iannelli, M., "Mathematical problems in the description of age structured populations", *Proc. Internat. Conf. Bari-Italy* (1983), 18–22.

[52] Iannelli, M., *Mathematical Theory of Age-Structured Population Dynamics*, CNR Applied Mathematics Monographs, Vol. 7, Giardini Editori e

Stampatori, Pisa (1995).

[53] Impagliazzo, J., *Deterministic Aspects of Mathematical Demography*, Biomathematics, Vol. 13, Springer-Verlag, Berlin (1985).

[54] Keyfitz, N., *Introduction to the Mathematical Populations with Revisions*, Addison-Wesley (1977).

[55] Keyfitz, N., *Applied Mathematical Demography*, Springer-Verlag, New York (1985).

[56] Kubo, M., "Subdifferential operator approach to nonlinear age dependent population dynamics", *Japan J. Appl. Math.* **5** (1988), 225–256.

[57] Kubo, M., and Langlais, M., "Periodic solutions for a population dynamics problem with age-dependence and spatial structure", *J. Math. Biol.* **29** (1991), 363–378.

[58] Ladyzenskaya, O.A., Solonnikov, V.A. and Uraltzeva, N.N., *Linear and Quasilinear Equations of Parabolic Type*, Nauka, Moskow (1967).

[59] Langlais, M., "A nonlinear problem in age dependent population diffusion", *SIAM J. Math. Anal.* **16** (1985), 510–529.

[60] Langlais, M., "Large time behaviour in a nonlinear age-dependent population dynamics problem with spatial diffusion", *J. Math. Biol.* **26** (1988), 319–346.

[61] Langhaar, H., "General population theory in age-time continuum", *J. Franklin Inst.* **293** (1972), 199–214.

[62] Lebeau, G., and Robbiano, L., "Contrôle exact de l'équation de la chaleur", *Comm. P.D.E.* **30** (1995), 335–357.

[63] Lefkovitch, L., "The study of population growth in organisms grouped by stages", *Biometrics* **21** (1965), 1–18.

[64] Lefkovitch, L., "A theoretical evaluation of population growth after removing individuals from some age groups", *Bull. Ent. Res.* **57** (1966), 437–445.

[65] Leung, A., *Systems of Nonlinear Partial Differential Equations, Applications to Biology and Engineering*, Kluwer Academic Publishers, Boston (1989).

[66] Lions, J.L., *Contrôle des systèmes distribués singuliers*, MMI 13, Gauthier Villars (1983).

[67] Lotka, A.J., "The stability of the normal age distribution", *Proc. Nat. Acad. Sci.* **8** (1922), 339–345.

[68] Malthus, T.R., *An Essay on the Principle of Population*, (First edition), London (1798).

[69] Marcati, P., "Asymptotic behaviour in age dependent population diffusion model", *SIAM J. Math. Anal.* **12** (1981), 904–916.

[70] McKendrick, A.C., "Applications of mathematics to medical problems", *Proc. Edinburgh Math. Soc.* **44** (1926), 98–130.

[71] Medhin, N., "Optimal harvesting in age-structured population", *J. Optim. Theory. Appl.* **74 (3)** (1992), 413–423.

[72] Metz, J.A.J., and Diekmann, O., *The Dynamics of Physiologically Structured Populations*, Springer Lecture Notes in Biomathematics 68 (1986).

[73] Mizohata, S., "Unicité du prolongement des solutions pour quelques opérateurs différentiels paraboliques", *Mem. Coll. Sc. Univ. Kyoto* **31** (1958), 219–239.

[74] Mueller, J., "Optimal vaccination strategies - for whom ?", *Math. Biosci.* **139** (1997), 133–154.

[75] Murphy, L.F., and Smith, S.J., "Optimal harvesting of an age-structured population", *J. Math. Biol.* **29** (1990), 77–90.

[76] Murray, J.D., *Mathematical Biology*, Springer-Verlag, Berlin (1989).

[77] Rockafellar, R.T., *Convex Analysis*, Princeton University Press (1970).

[78] Rores, C., and Fair, W., "Optimal age specific harvesting policy for continuous-time population model", in *Modeling and Differential Equations in Biology*, T.A. Burton (Ed.), M. Dekker, New York (1980).

[79] Schenzle, D., "An age-structured model for pre and post-vaccination measles transmission", *IMA J. Math. Appl. Biol. Med.* **1** (1984), 169–191.

[80] Sharpe, F.R., and Lotka, A.J., "A problem in age-distribution", *Philosophical Magazine* **21** (1911), 435–438.

[81] Song, J., and Yu, J., *Population System Control*, Springer-Verlag, New York (1977).

[82] Thieme, H., "Renewal theorems for linear periodic Volterra integral equations", *J. Int. Eqs.* **7** (1984), 253–277.

[83] Verhulst, P.F., "A note on the law of population growth", Correspondence Mathématique et Physique Publiée par A. Quetelet, Vol. 10, Brussels (1838).

[84] Vrabie, I., *Compactness Methods for Nonlinear Evolutions*, Longman (1995).

[85] Webb, G., "Nonlinear semigroups and age-dependent population models", *Ann. Mat. Pura Appl.* **4** (1981), 43–55.

[86] Webb, G., "A semigroup proof of the Sharpe-Lotka theorem", in *Infinite Dimensional System*, F. Kappel and W. Schappacher (Eds.), Lecture Notes in Mathematics 1076, Springer-Verlag, Berlin (1984), 254–268.

[87] Webb, G., *Theory of Nonlinear Age-Dependent Population Dynamics*, Marcel Dekker, New York (1985).

INDEX

MATHEMATICAL MODELLING:
Theory and Applications

---

1. M. Křížek and P. Neittaanmäki: *Mathematical and Numerical Modelling in Electrical Engineering.* Theory and Applications. 1996
ISBN 0-7923-4249-6
2. M.A. van Wyk and W.-H. Steeb: *Chaos in Electronics.* 1997
ISBN 0-7923-4576-2
3. A. Halanay and J. Samuel: *Differential Equations, Discrete Systems and Control.* Economic Models. 1997
ISBN 0-7923-4675-0
4. N. Meskens and M. Roubens (eds.): *Advances in Decision Analysis.* 1999
ISBN 0-7923-5563-6
5. R.J.M.M. Does, K.C.B. Roes and A. Trip: *Statistical Process Control in Industry.* Implementation and Assurance of SPC. 1999
ISBN 0-7923-5570-9
6. J. Caldwell and Y.M. Ram: *Mathematical Modelling.* Concepts and Case Studies. 1999
ISBN 0-7923-5820-1
7. 1. R. Haber and L. Keviczky: *Nonlinear System Identification - Input-Output Modeling Approach.* Volume 1: Nonlinear System Parameter Identification. 1999
ISBN 0-7923-5856-2; ISBN 0-7923-5858-9 Set
   2. R. Haber and L. Keviczky: *Nonlinear System Identification - Input-Output Modeling Approach.* Volume 2: Nonlinear System Structure Identification. 1999
ISBN 0-7923-5857-0; ISBN 0-7923-5858-9 Set
8. M.C. Bustos, F. Concha, R. Bürger and E.M. Tory: *Sedimentation and Thickening.* Phenomenological Foundation and Mathematical Theory. 1999
ISBN 0-7923-5960-7
9. A.P. Wierzbicki, M. Makowski and J. Wessels (eds.): *Model-Based Decision Support Methodology with Environmental Applications.* 2000
ISBN 0-7923-6327-2
10. C. Rocşoreanu, A. Georgescu and N. Giurgiţeanu: *The FitzHugh-Nagumo Model.* Bifurcation and Dynamics. 2000
ISBN 0-7923-6427-9
11. S. Aniţa: *Analysis and Control of Age-Dependent Population Dynamics.* 2000
ISBN 0-7923-6639-5

KLUWER ACADEMIC PUBLISHERS – DORDRECHT / BOSTON / LONDON

MIX
Papier aus verantwortungsvollen Quellen
Paper from responsible sources
FSC® C105338

If you have any concerns about our products,
you can contact us on
ProductSafety@springernature.com

In case Publisher is established outside the EU,
the EU authorized representative is:
Springer Nature Customer Service Center GmbH
Europaplatz 3, 69115 Heidelberg, Germany

Printed by Libri Plureos GmbH
in Hamburg, Germany